Numerical Methods and Modeling for Chemical Engineers

Numerical Methods and Modeling for Chemical Engineers

Mark E. Davis
Virginia Polytechnic Institute and State University

John Wiley & Sons
New York Chichester Brisbane Toronto Singapore

Library of Congress Cataloging in Publication Data:

Davis, Mark E.
Numerical methods and modeling for chemical
engineers.

Bibliography: p.
Includes index.

1. Chemical engineering—Mathematical models.
2. Differential equations. I. Title.

TP155.D33 1984 660.2′8′0724 83-21590
ISBN 0-471-88761-7

Printed in the United States of America

10 9 8 7 6 5 4 3 2 1

To Mary Margaret

Preface

This book is an introduction to the quantitative treatment of differential equations that arise from modeling physical phenomena in the area of chemical engineering. It evolved from a set of notes developed for courses taught at Virginia Polytechnic Institue and State University.

An engineer working on a mathematical project is typically not interested in sophisticated theoretical treatments, but rather in the solution of a model and the physical insight that the solution can give. A recent and important tool in regard to this objective is mathematical software—preprogrammed, reliable computer subroutines for solving mathematical problems. Since numerical methods are not infallible, a "black-box" approach of using these subroutines can be dangerous. To utilize software effectively, one must be aware of its capabilities and especially its limitations. This implies that the user must have at least an intuitive understanding of how the software is designed and implemented. Thus, although the subjects covered in this book are the same as in other texts, the treatment is different in that it emphasizes the methods implemented in commercial software. The aim is to provide an understanding of how the subroutines work in order to help the engineer gain maximum benefit from them.

This book outlines numerical techniques for differential equations that either illustrate a computational property of interest or are the underlying methods of a computer software package. The intent is to provide the reader with sufficient background to effectively utilize mathematical software. The reader is assumed to have a basic knowledge of mathematics, and results that require extensive mathematical literacy are stated with proper references. Those who desire to delve deeper into a particular subject can then follow the leads given in the references and bibliographies.

Each chapter is provided with examples that further elaborate on the text. Problems at the end of each chapter are aimed at mimicking industrial mathematics projects and, when possible, are extensions of the examples in the text. These problems have been grouped into two classes:

Class 1: Problems that illustrate direct numerical application of the formulas in the text.

Class 2: Problems that should be solved with software of the type described in the text (designated by an asterisk after the problem number).

The level of this book is introductory, although the latest techniques are presented. The book can serve as a text for a senior or first-year graduate level course. At Virginia Polytechnic Institute and State University I have successfully used this material for a two-quarter sequence of first-year graduate courses. In the first quarter ordinary differential equations, Chapter 1 to 3, are covered. The second quarter examines partial differential equations using Chapters 4 and 5.

I gratefully acknowledge the following individuals who have either directly or indirectly contributed to this book: Kenneth Denison, Julio Diaz, Peter Mercure, Kathleen Richter, Peter Rony, Layne Watson, and John Yamanis. I am especially indebted to Graeme Fairweather who read the manuscript and provided many helpful suggestions for its improvement. I also thank the Department of Chemical Engineering at Virginia Polytechnic Institute and State University for its support, and I apologize to the many graduate students who suffered through the early drafts as course texts. Last, and most of all, my sincerest thanks go to Jan Chance for typing the manuscript in her usual flawless form.

I dedicate this book to my wife, who uncomplainingly gave up a portion of her life for its completion.

Mark E. Davis

Contents

Chapter 2 Boundary-Value Problems for Ordinary Differential Equations: Discrete Variable Methods 53

Chapter 3 Boundary-Value Problems for Ordinary Differential Equations: Finite Element Methods 97

Numerical Methods and Modeling for Chemical Engineers

1

Initial-Value Problems for Ordinary Differential Equations

INTRODUCTION

The goal of this book is to expose the reader to modern computational tools for solving differential equation models that arise in chemical engineering, e.g., diffusion-reaction, mass-heat transfer, and fluid flow. The emphasis is placed on the understanding and proper use of software packages. In each chapter we outline numerical techniques that either illustrate a computational property of interest or are the underlying methods of a computer package. At the close of each chapter a survey of computer packages is accompanied by examples of their use.

BACKGROUND

Many problems in engineering and science can be formulated in terms of differential equations. A differential equation is an equation involving a relation between an unknown function and one or more of its derivatives. Equations involving derivatives of only one independent variable are called ordinary differential equations and may be classified as either initial-value problems (IVP) or boundary-value problems (BVP). Examples of the two types are:

$$\text{IVP:} \quad y'' = -yx \tag{1.1a}$$

$$y(0) = 2, \qquad y'(0) = 1 \tag{1.1b}$$

$$\text{BVP:} \quad y'' = -yx \tag{1.2a}$$

$$y(0) = 2, \qquad y(1) = 1 \tag{1.2b}$$

where the prime denotes differentiation with respect to x. The distinction between the two classifications lies in the location where the extra conditions [Eqs. (1.1b) and (1.2b)] are specified. For an IVP, the conditions are given at the same value of x, whereas in the case of the BVP, they are prescribed at two different values of x.

Since there are relatively few differential equations arising from practical problems for which analytical solutions are known, one must resort to numerical methods. In this situation it turns out that the numerical methods for each type of problem, IVP or BVP, are quite different and require separate treatment. In this chapter we discuss IVPs, leaving BVPs to Chapters 2 and 3.

Consider the problem of solving the mth-order differential equation

$$y^{(m)} = f(x, y, y', y'', \ldots, y^{(m-1)}) \tag{1.3}$$

with initial conditions

$$\begin{aligned} y(x_0) &= y_0 \\ y'(x_0) &= y'_0 \\ &\vdots \\ y^{(m-1)}(x_0) &= y_0^{(m-1)} \end{aligned}$$

where f is a known function and $y_0, y'_0, \ldots, y_0^{(m-1)}$ are constants. It is customary to rewrite (1.3) as an equivalent system of m first-order equations. To do so, we define a new set of dependent variables $y_1(x), y_2(x), \ldots, y_m(x)$ by

$$\begin{aligned} y_1 &= y \\ y_2 &= y' \\ y_3 &= y'' \\ &\vdots \\ y_m &= y^{(m-1)} \end{aligned} \tag{1.4}$$

and transform (1.3) into

$$\begin{aligned} y'_1 &= y_2 && = f_1(x, y_1, y_2, \ldots, y_m) \\ y'_2 &= y_3 && = f_2(x, y_1, y_2, \ldots, y_m) \\ &\vdots && \vdots \\ y'_m &= f(x, y_1, y_2, \ldots, y_m) && = f_m(x, y_1, y_2, \ldots, y_m) \end{aligned} \tag{1.5}$$

with

$$\begin{aligned} y_1(x_0) &= y_0 \\ y_2(x_0) &= y'_0 \\ &\vdots \\ y_m(x_0) &= y_0^{(m-1)} \end{aligned}$$

In vector notation (1.5) becomes

$$\mathbf{y}'(x) = \mathbf{f}(x, \mathbf{y}) \qquad \textbf{(1.6)}$$
$$\mathbf{y}(x_0) = \mathbf{y}_0$$

where

$$\mathbf{y}(x) = \begin{bmatrix} y_1(x) \\ y_2(x) \\ \vdots \\ y_m(x) \end{bmatrix}, \qquad \mathbf{f}(x, \mathbf{y}) = \begin{bmatrix} f_1(x, \mathbf{y}) \\ f_2(x, \mathbf{y}) \\ \vdots \\ f_m(x, \mathbf{y}) \end{bmatrix}, \qquad \mathbf{y}_0 = \begin{bmatrix} y_0 \\ y_0' \\ \vdots \\ y_0^{(m-1)} \end{bmatrix}$$

It is easy to see that (1.6) can represent either an mth-order differential equation, a system of equations of mixed order but with total order of m, or a system of m first-order equations. In general, subroutines for solving IVPs assume that the problem is in the form (1.6). In order to simplify the analysis, we begin by examining a single first-order IVP, after which we extend the discussion to include systems of the form (1.6).

Consider the initial-value problem

$$y' = f(x, y), \qquad y(x_0) = y_0 \qquad \textbf{(1.7)}$$
$$x_0 \leqslant x \leqslant x_N$$

We assume that $\partial f/\partial y$ is continuous on the strip $x_0 \leqslant x \leqslant x_N$, thus guaranteeing that (1.7) possesses a unique solution [1]. If $y(x)$ is the exact solution to (1.7), its graph is a curve in the xy-plane passing through the point (x_0, y_0). A discrete numerical solution of (1.7) is defined to be a set of points $[(x_i, u_i)]_{i=0}^N$, where $u_0 = y_0$ and each point (x_i, u_i) is an approximation to the corresponding point $(x_i, y(x_i))$ on the solution curve. Note that the numerical solution is only a set of points, and nothing is said about values between the points. In the remainder of this chapter we describe various methods for obtaining a numerical solution $[(x_i, u_i)]_{i=0}^N$.

EXPLICIT METHODS

We again consider (1.7) as the model differential equation and begin by dividing the interval $[x_0, x_N]$ into N equally spaced subintervals such that

$$h = \frac{x_N - x_0}{N} \qquad \textbf{(1.8)}$$
$$x_i = x_0 + ih, \qquad i = 0, 1, 2, \ldots, N$$

The parameter h is called the step-size and does not necessarily have to be uniform over the interval. (Variable step-sizes are considered later.)

If $y(x)$ is the exact solution of (1.7), then by expanding $y(x)$ about the point x_i using Taylor's theorem with remainder we obtain:

$$y(x_{i+1}) = y(x_i) + (x_{i+1} - x_i)y'(x_i) + \frac{(x_{i+1} - x_i)^2}{2!} y''(\xi_i), \qquad x_i \leqslant \xi_i \leqslant x_{i+1} \tag{1.9}$$

The substitution of (1.7) into (1.9) gives

$$y(x_{i+1}) = y(x_i) + hf(x_i, y(x_i)) + \frac{h^2}{2!} f'(\xi_i, y(\xi_i)) \tag{1.10}$$

The simplest numerical method is obtained by truncating (1.10) after the second term. Thus with $u_i \simeq y(x_i)$,

$$\begin{aligned} u_{i+1} &= u_i + hf(x_i, u_i), \qquad i = 0, 1, \ldots, N - 1, \\ u_0 &= y_0 \end{aligned} \tag{1.11}$$

This method is called the Euler method.

By assuming that the value of u_i is exact, we find that the application of (1.11) to compute u_{i+1} creates an error in the value of u_{i+1}. This error is called the local truncation error, e_{i+1}. Define the local solution, $z(x)$, by

$$z'(x) = f(x, z), \qquad z(x_i) = u_i \tag{1.12}$$

An expression for the local truncation error, $e_{i+1} = z(x_{i+1}) - u_{i+1}$, can be obtained by comparing the formula for u_{i+1} with the Taylor's series expansion of the local solution about the point x_i. Since

$$z(x_i + h) = z(x_i) + hf(x_i, z(x_i)) + \frac{h^2}{2!} z''(\bar{\xi}_i)$$

or

$$z(x_i + h) = u_i + hf(x_i, u_i) + \frac{h^2}{2!} z''(\bar{\xi}_i), \qquad x_i \leqslant \bar{\xi}_i \leqslant x_{i+1} \tag{1.13}$$

it follows that

$$e_{i+1} = \frac{h^2}{2!} z''(\bar{\xi}_i) = 0(h^2) \tag{1.14}$$

The notation 0( ) denotes terms of order ( ), i.e., $f(h) = 0(h^L)$ if $|f(h)| \leqslant Ah^l$ as $h \to 0$, where A and l are constants [1]. The global error is defined as

$$\mathscr{E}_{i+1} = y(x_{i+1}) - u_{i+1} \tag{1.15}$$

and is thus the difference between the true solution and the numerical solution at $x = x_{i+1}$. Notice the distinction between e_{i+1} and $\mathscr{E}_{i+1}$. The relationships between e_{i+1} and $\mathscr{E}_{i+1}$ will be discussed later in the chapter.

We say that a method is pth-order accurate if

$$e_{i+1} = 0(h^{p+1}) \tag{1.16}$$

and from (1.14) and (1.16) the Euler method is first-order accurate. From the previous discussions one can see that the local truncation error in each step can be made as small as one wishes provided the step-size is chosen sufficiently small.

The Euler method is explicit since the function f is evaluated with known information (i.e., at the left-hand side of the subinterval). The method is pictured in Figure 1.1. The question now arises as to whether the Euler method is able to provide an accurate approximation to (1.7). To partially answer this question, we consider Example 1, which illustrates the properties of the Euler method.

EXAMPLE 1

Kehoe and Butt [2] have studied the kinetics of benzene hydrogenation on a supported Ni/kieselguhr catalyst. In the presence of a large excess of hydrogen, the reaction is pseudo-first-order at temperatures below 200°C with the rate given by

$$-r = P_{H_2} k_0 K_0 T \exp\left[\frac{(-Q - E_a)}{R_g T}\right] C_B \qquad \text{mole/(g of catalyst·s)}$$

where

$$\begin{aligned} R_g &= \text{gas constant, 1.987 cal/(mole·K)} \\ -Q - E_a &= \text{2700 cal/mole} \\ P_{H_2} &= \text{hydrogen partial pressure (torr)} \\ k_0 &= \text{4.22 mole/(gcat·s·torr)} \\ K_0 &= 2.63 \times 10^{-6}\ \text{cm}^3\text{/(mole·K)} \\ T &= \text{absolute temperature (K)} \\ C_B &= \text{concentration of benzene (mole/cm}^3\text{).} \end{aligned}$$

Price and Butt [3] studied this reaction in a tubular reactor. If the reactor is assumed to be isothermal, we can calculate the dimensionless concentration profile of benzene in their reactor given plug flow operation in the absence of inter- and intraphase gradients. Using a typical run,

$$\begin{aligned} P_{H_2} &= \text{685 torr} \\ \rho_B &= \text{density of the reactor bed, 1.2 gcat/cm}^3 \\ \theta &= \text{contact time, 0.226 s} \\ T &= 150°\text{C} \end{aligned}$$

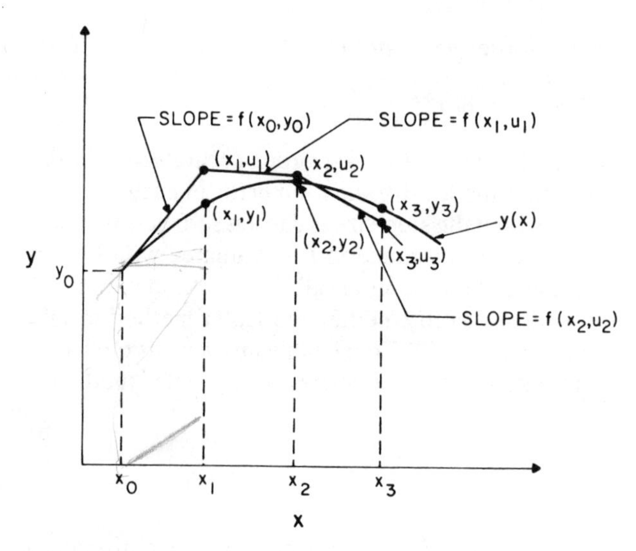

FIGURE 1.1 Euler method.

SOLUTION

Define

C_B^0 = feed concentration of benzene (mole/cm^3)

z = axial reactor coordinate (cm)

L = reactor length

y = dimensionless concentration of benzene (C_B/C_B^0)

x = dimensionless axial coordinate (z/L).

The one-dimensional steady-state material balance for the reactor that expresses the fact that the change in the axial convection of benzene is equal to the amount converted by reaction is

$$\frac{d}{dx}\left(\frac{C_B}{\theta}\right) = r$$

with

$$C_B = C_B^0 \quad \text{at} \quad x = 0$$

Since θ is constant,

$$\frac{dy}{dx} = -\rho_B \theta P_{H_2} k_0 K_0 T \exp\left[\frac{(-Q - E_a)}{R_g T}\right] y$$

Let

$$\phi = \rho_B \theta P_{H_2} k_0 K_0 T \exp\left[\frac{(-Q - E_a)}{R_g T}\right]$$

Using the data provided, we have $\phi = 21.6$. Therefore, the material balance equation becomes

$$\frac{dy}{dx} = -21.6y$$

with

$$y = 1 \quad \text{at} \quad x = 0$$

and analytical solution

$$y = \exp(-21.6x)$$

Now we solve the material balance equation using the Euler method [Eq. (1.11)]:

$$u_{i+1} = u_i - 21.6hu_i, \qquad i = 0, 1, 2, \ldots, N - 1$$

where

$$h = \frac{1}{N}$$

Table 1.1 shows the generated results. Notice that for $N = 10$ the differences between the analytical solution and the numerical approximation increase with x. In a problem where the analytical solution decreases with increasing values of the independent variable, a numerical method is unstable if the global error grows with increasing values of the independent variable (for a rigorous definition of stability, see [4]). Therefore, for this problem the Euler method is unstable when $N = 10$. For $N = 20$ the global error decreases with x, but the solution oscillates in sign. If the error decreases with increasing x, the method is said to be stable. Thus with $N = 20$ the Euler method is stable (for this problem), but the solution contains oscillations. For all $N > 20$, the method is stable and produces no oscillations in the solution.

From a practical standpoint, the "effective" reaction zone would be approximately $0 \leq x \leq 0.2$. If the reactor length is reduced to $0.2L$, then a more realistic problem is produced. The material balance equation becomes

$$\frac{dy}{dx} = -4.32y$$

$$y = 1 \quad \text{at} \quad x = 0$$

Results for the "short" reactor are shown in Table 1.2. As with Table 1.1, we see that a large number of steps are required to achieve a "good" approximation to the analytical solution. An explanation of the observed behavior is provided in the next section.

Physically, the solutions are easily rationalized. Since benzene is a reactant, thus being converted to products as the fluid progresses toward the reactor outlet ($x = 1$), y should decrease with x. Also, a longer reactor would allow for greater conversion, i.e., smaller y values at $x = 1$.

TABLE 1.1 Results of Euler Method on $\frac{dy}{dx} = -21.6y$, $y = 1$ at $x = 0$

x	Analytical Solution†	$N = 10$	$N = 20$	$N = 100$	$N = 8000$
0.00	1.00000	1.0000	1.00000	1.00000	1.00000
0.05	0.33960	—	−0.80000(−1)	0.29620	0.33910
0.10	0.11533	−1.1600	0.64000(−2)	0.87733(−1)	0.11499
0.15	0.39164(−1)	—	−0.51200(−3)	0.25986(−1)	0.38993(−1)
0.20	0.13300(−1)	1.3456	0.40960(−4)	0.76970(−2)	0.13222(−1)
0.25	0.45166(−2)	—	−0.32768(−5)	0.22798(−2)	0.44837(−2)
0.30	0.15338(−2)	−1.5609	0.26214(−6)	0.67528(−3)	0.15204(−2)
0.35	0.52088(−3)	—	−0.20972(−7)	0.20000(−3)	0.51558(−3)
0.40	0.17689(−3)	1.8106	0.16777(−8)	0.59244(−4)	0.17483(−3)
0.45	0.60070(−4)	—	−0.13422(−9)	0.17548(−4)	0.59286(−4)
0.50	0.20400(−4)	−2.1003	0.10737(−10)	0.51976(−5)	0.20104(−4)
0.55	0.69276(−5)	—	−0.85899(−12)	0.15395(−5)	0.68172(−5)
0.60	0.23526(−5)	2.4364	0.68719(−13)	0.45600(−6)	0.23117(−5)
0.65	0.79892(−6)	—	−0.54976(−14)	0.13507(−6)	0.78390(−6)
0.70	0.27131(−6)	−2.8262	0.43980(−15)	0.40006(−7)	0.26582(−6)
0.75	0.92136(−7)	—	−0.35184(−16)	0.11850(−7)	0.90139(−7)
0.80	0.31289(−7)	3.2784	0.28147(−17)	0.35098(−8)	0.30566(−7)
0.85	0.10626(−7)	—	−0.22518(−18)	0.10396(−8)	0.10365(−7)
0.90	0.36084(−8)	−3.8030	0.18014(−19)	0.30793(−9)	0.35148(−8)
0.95	0.12254(−8)	—	−0.14412(−20)	0.91207(−10)	0.11919(−8)
1.00	0.41614(−9)	4.4114	0.11529(−21)	0.27015(−10)	0.40416(−9)

† (−3) denotes 1.0×10^{-3}.

STABILITY

In Example 1 it was seen that for some choices of the step-size, the approximate solution was unstable, or stable with oscillations. To see why this happens, we will examine the question of stability using the test equation

$$\frac{dy}{dx} = \lambda y \qquad y(0) = y_0 \tag{1.17}$$

where λ is a complex constant. Application of the Euler method to (1.17) gives

$$u_{i+1} = u_i + \lambda h u_i \tag{1.18}$$

or

$$u_{i+1} = (1 + h\lambda)u_i = (1 + h\lambda)^2 u_{i-1} = \ldots = (1 + h\lambda)^{i+1} u_0 \tag{1.19}$$

The analytical solution of (1.17) is

$$y(x_{i+1}) = y_0 e^{\lambda x_{i+1}} = y_0 e^{(i+1)h\lambda} \tag{1.20}$$

Comparing (1.20) with (1.19) shows that the application of Euler's method to (1.17) is equivalent to using the expression $(1 + h\lambda)$ as an approximation for

TABLE 1.2 Results of Euler Method on $\frac{dy}{dx} = -4.32y$, $y = 1$ at $x = 0$

x	Analytical Solution	$N = 100$	$N = 1000$	$N = 8000$
0.0	1.00000	1.00000	1.00000	1.00000
0.1	0.64921	0.64300	0.64860	0.64913
0.2	0.42147	0.41345	0.42068	0.42137
0.3	0.27362	0.26585	0.27286	0.27353
0.4	0.17764	0.17094	0.17698	0.17756
0.5	0.11533	0.10992	0.11479	0.11526
0.6	0.07487	0.07067	0.07445	0.07481
0.7	0.04860	0.04544	0.04828	0.04856
0.8	0.03155	0.02922	0.03132	0.03152
0.9	0.02048	0.01878	0.02031	0.02046
1.0	0.01330	0.01208	0.01317	0.01328

$e^{\lambda h}$. Now suppose that the value y_0 is not exactly representable by a machine number (see Appendix A), then $e_0 = y_0 - u_0$ will be nonzero. From (1.19), with u_0 replaced by $y_0 - e_0$,

$$u_{i+1} = (1 + h\lambda)^{i+1} (y_0 - e_0)$$

and the global error $\mathcal{E}_{i+1}$ is

$$\mathcal{E}_{i+1} = y(x_{i+1}) - u_{i+1} = y_0 e^{(i+1)h\lambda} - (1 + h\lambda)^{i+1} (y_0 - e_0)$$

or

$$\mathcal{E}_{i+1} = [e^{(i+1)\lambda h} - (1 + h\lambda)^{i+1}] \, y_0 + (1 + h\lambda)^{i+1} e_0 \tag{1.21}$$

Hence, the global error consists of two parts. First, there is an error that results from the Euler method approximation $(1 + h\lambda)$ for $e^{\lambda h}$. The second part is the propagation effect of the initial error, e_0. Clearly, if $|1 + h\lambda| > 1$, this component will grow and, no matter what the magnitude of e_0 is, it will become the dominant term in $\mathcal{E}_{i+1}$. Therefore, to keep the propagation effects of previous errors bounded when using the Euler method, we require

$$|1 + h\lambda| \leq 1 \tag{1.22}$$

The region of absolute stability is defined by the set of h (real nonnegative) and λ values for which a perturbation in a single value u_i will produce a change in subsequent values that does not increase from step to step [4]. Thus, one can see from (1.22) that the stability region for (1.17) corresponds to a unit disk in the complex $h\lambda$-plane centered at $(-1, 0)$. If λ is real, then

$$-2 \leq h\lambda \leq 0 \tag{1.23}$$

Notice that if the propagation effect is the dominant term in (1.21), the global error will oscillate in sign if $-2 \leq h\lambda \leq -1$.

EXAMPLE 2

Referring to Example 1, find the maximum allowable step-size for stability and for nonoscillatory behavior for the material balance equations of the "long" and "short" reactor. Can you now explain the behavior shown in Tables 1.1 and 1.2?

SOLUTION

For the long reactor: $\lambda_L = -21.6$ (real)
For the short reactor: $\lambda_S = -4.32$ (real)
For stability: $0 \geq h\lambda \geq -2$
For nonoscillatory error: $0 \geq h\lambda > -1$

	Long Reactor	Short Reactor
Unstable	$0.0926 < h$	$0.4630 < h$
Stable, error oscillations	$0.0463 \leq h \leq 0.0926$	$0.2315 \leq h \leq 0.4630$
Stable, no error oscillations	$h < 0.0463$	$h < 0.2315$

For the short reactor, all of the presented solutions are stable and nonoscillatory since the step-size is always less than 0.2315. The large number of steps required for a "reasonably" accurate solution is a consequence of the first-order accuracy of the Euler method.

For the long reactor with $N > 20$ the solutions are stable and nonoscillatory since h is less than 0.0463. With $N = 10$, $h = 0.1$ and the solution is unstable, while for $N = 20$, $h = 0.05$ and the solution is stable and oscillatory. From the above table, when $N = 20$, the global error should oscillate if the propagation error is the dominant term in Eq. (1.21). This behavior is not observed from the results shown in Table 1.1. The data for $N = 10$ and $N = 20$ can be explained by examining Eq. (1.21):

$$\mathscr{E}_{i+1} = [e^{(i+1)\lambda h} - (1 + h\lambda)^{i+1}]y_0 + (1 + h\lambda)^{i+1}e_0 = (\text{A})y_0 + (\text{B})e_0$$

For $N = 10$, $h = 0.1$ and $\lambda h = -2.16$. Therefore,

i	(A)	(B)	Global Error Calculated from Results Shown in Table 1.1
0	1.2753	−1.160	1.2753
1	−1.3323	1.3456	−1.3323
2	1.5624	−1.5609	1.5624

Since $y_0 = 1$ and e_0 is small, the global error is dominated by term (A) and not the propagation term, i.e., term (B). For $N = 20$, $h = 0.05$ and $\lambda h = -1.08$. Therefore,

i	(A)	(B)	Global Error Calculated from Results Shown in Table 1.1
0	0.4196	-0.08	0.4196
1	0.1089	0.64×10^{-2}	0.1089
2	0.3967×10^{-1}	-0.512×10^{-3}	0.3967×10^{-1}

As with $N = 10$, the global error is dominated by the term (A). Thus no oscillations in the global error are seen for $N = 20$.

From (1.19) one can explain the oscillations in the solution for $N = 10$ and 20. If

$$(1 + h\lambda) < 0$$

then the numerical solution will alternate in sign. For $(1 + h\lambda)$ to be equal to zero, $h\lambda = -1$. When $N = 10$ or 20, $h\lambda$ is less than -1 and therefore oscillations in the solution occur.

For this problem, it was shown that for the long reactor with $N = 10$ or 20 the propagation error was not the dominant part of the global error. This behavior is a function of the parameter λ and thus will vary from problem to problem.

From Examples 1 and 2 one observes that there are two properties of the Euler method that could stand improvement: stability and accuracy. Implicit within these categories is the cost of computation. Since the step-size of the Euler method has strict size requirements for stability and accuracy, a large number of function evaluations are required, thus increasing the cost of computation. Each of these considerations will be discussed further in the following sections. In the next section we will show methods that improve the order of the accuracy.

RUNGE-KUTTA METHODS

Runge-Kutta methods are explicit algorithms that involve evaluation of the function f at points between x_i and x_{i+1}. The general formula can be written as

$$u_{i+1} = u_i + \sum_{j=1}^{\nu} \omega_j K_j \tag{1.24}$$

where

$$\begin{aligned} K_j &= hf\left(x_i + c_j h,\ u_i + \sum_{l=1}^{j-1} a_{jl} K_l\right) \\ c_1 &= 0 \end{aligned} \tag{1.25}$$

Notice that if $\nu = 1$, $\omega_1 = 1$, and $K_1 = hf(x_i, u_i)$, the Euler method is obtained. Thus, the Euler method is the lowest-order Runge-Kutta formula. For higher-order formulas, the parameters ω, c, and a are found as follows. For example, if $\nu = 2$, first expand the exact solution of (1.7) in a Taylor's series,

$$y(x_{i+1}) = y(x_i) + hf(x_i, y(x_i)) + \frac{h^2}{2!} f'(x_i, y(x_i)) + 0(h^3) \qquad \textbf{(1.26)}$$

Next, rewrite $f'(x_i, y(x_i))$ as

$$\frac{df_i}{dx} = \frac{\partial f_i}{\partial x} + \frac{\partial f_i}{\partial y} \frac{dy}{dx}\bigg|_{x=x_i} = (f_x + f_y f)_i \qquad \textbf{(1.27)}$$

Substitute (1.27) into (1.26) and truncate the $0(h^3)$ term to give

$$u_{i+1} = u_i + hf_i + \frac{h^2}{2} (f_x + f_y f)_i \qquad \textbf{(1.28)}$$

Expand each of the K_j's about the ith position. To do so, denote

$$K_1 = hf(x_i, u_i) = hf_i \qquad \textbf{(1.29a)}$$

and

$$K_2 = hf(x_i + c_2 h, u_i + a_{21} K_1) \qquad \textbf{(1.29b)}$$

Recall that for any two functions η and ϕ that are located near x_i and u_i, respectively,

$$f(\eta, \phi) \simeq f(x_i, u_i) + (\eta - x_i) f_x(x_i, u_i) + (\phi - u_i) f_y(x_i, u_i) \qquad \textbf{(1.30)}$$

Using (1.30) on K_2 gives

$$K_2 = h(f_i + c_2 h f_x + a_{21} K_1 f_y)$$

or

$$K_2 = hf_i + h^2(c_2 f_x + a_{21} f_y f)_i \qquad \textbf{(1.31)}$$

Substitute (1.29a) and (1.31) into (1.24):

$$u_{i+1} = u_i + \omega_1 h f_i + \omega_2 h f_i + \omega_2 h^2 c_2 (f_x)_i + a_{21} \omega_2 h^2 (f_y f)_i \qquad \textbf{(1.32)}$$

Comparing like powers of h in (1.32) and (1.28) shows that

$$\omega_1 + \omega_2 = 1.0$$

$$\omega_2 c_2 = 0.5$$

$$\omega_2 a_{21} = 0.5$$

The Runge-Kutta algorithm is completed by choosing the free parameter; i.e., once either ω_1, ω_2, c_2, or a_{21} is chosen, the others are fixed by the above formulas.

If c_2 is set equal to 0.5, the Runge-Kutta scheme is

$$u_{i+1} = u_i + hf(x_i + \tfrac{1}{2}h,\ u_i + \tfrac{1}{2}hf_i), \qquad i = 0, 1, \ldots, N - 1 \tag{1.33}$$
$$u_0 = y_0$$

or a midpoint method. For $c_2 = 1$,

$$u_{i+1} = u_i + \frac{h}{2}[f_i + f(x_i + h,\ u_i + hf_i)], \qquad i = 0, 1, \ldots, N - 1 \tag{1.34}$$
$$u_0 = y_0$$

These two schemes are graphically interpreted in Figure 1.2. The methods are second-order accurate since (1.28) and (1.31) were truncated after terms of $0(h^2)$.

If a pth-order accurate formula is desired, one must take ν large enough so that a sufficient number of degrees of freedom (free parameters) are available in order to obtain agreement with a Taylor's series truncated after terms in h^p. A table of minimum such ν for a given p is

p	2	3	4	5	6	...
ν	2	3	4	6	8	...

Since ν represents the number of evaluations of the function f for a particular i, the above table shows the minimum amount of work required to achieve a desired order of accuracy. Notice that there is a jump in ν from 4 to 6 when p goes from 4 to 5, so traditionally, because of the extra work, methods with $p > \nu$ have been disregarded. An example of a fourth-order scheme is the

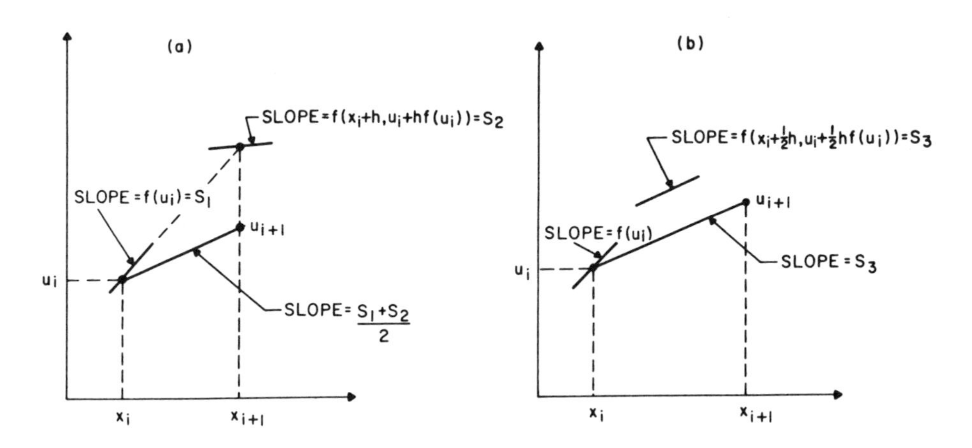

FIGURE 1.2 Runge-Kutta interpretations. (*a*) Eq. (1.34). (*b*) Eq. (1.33).

Runge-Kutta-Gill Method [41] and is:

$$u_{i+1} = u_i + \tfrac{1}{6}(K_1 + K_4) + \tfrac{1}{3}(bK_2 + dK_3)$$
$$K_1 = hf(x_i, u_i)$$
$$K_2 = hf(x_i + \tfrac{1}{2}h, u_i + \tfrac{1}{2}K_1)$$
$$K_3 = hf(x_i + \tfrac{1}{2}h, u_i + aK_1 + bK_2) \tag{1.35}$$
$$K_4 = hf(x_i + h, u_i + cK_2 + dK_3)$$
$$a = \frac{\sqrt{2} - 1}{2}, \qquad b = \frac{2 - \sqrt{2}}{2}$$
$$c = -\frac{\sqrt{2}}{2}, \qquad d = 1 + \frac{\sqrt{2}}{2}$$

for

$$i = 0, 1, \ldots, N - 1 \qquad \text{and} \qquad u_0 = y_0$$

The parameter choices in this algorithm have been made to minimize round-off error.

Use of the explicit Runge-Kutta formulas improves the order of accuracy, but what about the stability of these methods? For example, if λ is real, the second-order Runge-Kutta algorithm is stable for the region $-2.0 \leq \lambda h \leq 0$, while the fourth-order Runge-Kutta-Gill method is stable for the region $-2.8 \leq \lambda h \leq 0$.

EXAMPLE 3

A thermocouple at equilibrium with ambient air at 10°C is plunged into a warm-water bath at time equal to zero. The warm water acts as an infinite heat source at 20°C since its mass is many times that of the thermocouple. Calculate the response curve of the thermocouple.

Data: Time constant of the thermocouple = 0.4 min^{-1}.

SOLUTION

Define

C_p = thermal capacity of the thermocouple
U = heat transfer coefficient of the thermocouple
A = heat transfer area of thermocouple
t = time (min)
T, T_p, T_0 = temperature of thermocouple, water, and ambient air

$$\theta = \frac{T_p - T}{T_p - T_0}$$

$$\eta = \frac{C_p}{UA} = \text{time constant of the thermocouple}$$

$$t^* = \frac{t}{10}$$

The governing differential equation is Newton's law of heating or cooling and is

$$C_p \frac{dT}{dt} = UA(T_p - T), \qquad T = 10°\text{C} \quad \text{at} \quad t = 0$$

If the response curve is calculated for $0 \leqslant t \leqslant 10$ min, then

$$\frac{d\theta}{dt^*} = -25\theta, \qquad \theta = 1 \quad \text{at} \quad t = 0$$

The analytical solution is

$$\theta = e^{-25t^*}, \qquad 0 \leqslant t^* \leqslant 1$$

Now solve the differential equation using the second-order Runge-Kutta method [Eq. (1.34)]:

$$u_0 = 1$$

$$u_{i+1} = u_i + \frac{h}{2}[f_i + f(t_i^* + h, u_i + hf_i)], \qquad i = 0, 1, \ldots, N - 1$$

where

$$f_i = -25u_i$$

$$f(t_i^* + h, u_i + hf_i) = -25(u_i + hf_i)$$

and using the Runge-Kutta-Gill method [Eq. (1.35)]:

$$u_0 = 1$$

$$u_{i+1} = u_i + \tfrac{1}{6}(K_1 + K_4) + \tfrac{1}{3}(bK_2 + dK_3), \qquad i = 0, 1, \ldots, N - 1$$

$$K_1 = -25hu_i$$

$$K_2 = -25h(u_i + \tfrac{1}{2}K_1)$$

$$K_3 = -25h(u_i + aK_1 + bK_2)$$

$$K_4 = -25h(u_i + cK_2 + dK_3)$$

Table 1.3 shows the generated results. Notice that for $N = 20$ the second-order Runge-Kutta method shows large discrepancies when compared with the analytical solution. Since $\lambda = -25$, the maximum stable step-size for this method is $h = 0.08$, and for $N = 20$, h is very close to this maximum. For the

TABLE 1.3 Comparison of Runge-Kutta Methods $\frac{d\theta}{dt^*} = -25\theta$, $\theta = 1$ at $t^* = 0$

		Second-Order Runge-Kutta Method		Runge-Kutta-Gill Method	
t^*	Analytical Solution	$N = 20$	$N = 200$	$N = 20$	$N = 200$
0.00000	1.00000	1.00000	1.00000	1.00000	1.00000
0.20000	0.67379(−02)	0.79652(−01)	0.68350(−02)	0.89356(−02)	0.67380(−02)
0.40000	0.45400(−04)	0.63444(−02)	0.46717(−04)	0.79845(−04)	0.45401(−04)
0.60000	0.30590(−06)	0.50534(−03)	0.31931(−06)	0.71346(−06)	0.30591(−06)
0.80000	0.20612(−08)	0.40252(−04)	0.21825(−08)	0.63752(−08)	0.20612(−08)
1.00000	0.13888(−10)	0.32061(−05)	0.14917(−10)	0.56966(−10)	0.13889(−10)

Runge-Kutta-Gill method the maximum stable step-size is $h = 0.112$, and h never approaches this limit. From Table 1.3 one can also see that the Runge-Kutta-Gill method produces a more accurate solution than the second-order method, which is as expected since it is fourth-order accurate. To further this point, refer to Table 1.4 where we compare a first (Euler), a second, and a fourth-order method to the analytical solution. For a given N, the accuracy increases with the order of the method, as one would expect. Since the Runge-Kutta-Gill method (RKG) requires four function evaluations per step while the Euler method requires only one, which is computationally more efficient? One can answer this question by comparing the RKG results for $N = 100$ with the Euler results for $N = 800$. The RKG method ($N = 100$) takes 400 function evaluations to reach $t^* = 1$, while the Euler method ($N = 800$) takes 800. From Table 1.4 it can be seen that the RKG ($N = 100$) results are more accurate than the Euler ($N = 800$) results, and require half as many function evaluations. It is therefore shown that for this problem although more function evaluations per step are required by the higher-order accurate formulas, they are computationally more efficient when trying to meet a specified error tolerance (this result cannot be generalized to include all problems).

Physically, all the results in Tables 1.3 and 1.4 have significance. Since $\theta = (T_p - T)/(T_p - T_0)$, initially $T = T_0$ and $\theta = 1$. When the thermocouple is plunged into the water, the temperature will begin to rise and T will approach T_p, that is, θ will go to 0.

So far we have always illustrated the numerical methods with test problems that have an analytical solution so that the errors are easily recognizable. In a practical problem an analytical solution will not be known, so no comparisons can be made to find the errors occurring during computation. Alternative strategies must be constructed to estimate the error. One method of estimating the local error would be to calculate the difference between u_{i+1}^* and u_{i+1} where u_{i+1} is calculated using a step-size of h and u_{i+1}^* using a step-size of $h/2$. Since the accuracy of the numerical method depends upon the step-size to a certain power, u_{i+1}^* will be a better estimate for $y(x_{i+1})$ than u_{i+1}. Therefore,

$$|z_{i+1} - u_{i+1}^*| < |z_{i+1} - u_{i+1}|$$

TABLE 1.4 Comparison of Runge-Kutta Methods with the Euler Method

$$\frac{d\theta}{dt^*} = -25\theta, \theta = 1 \text{ at } t^* = 0$$

		Second-Order Runge-Kutta Method		Runge-Kutta-Gill Method		Euler Method	
t^*	Analytical Solution	$N = 100$	$N = 800$	$N = 100$	$N = 800$	$N = 100$	$N = 800$
0.00000	1.000000	1.00000	1.00000	1.00000	1.00000	1.00000	1.00000
0.20000	0.67379(−02)	0.71746(−02)	0.67436(−02)	0.67393(−02)	0.67379(−02)	0.31712(−02)	0.62212(−02)
0.40000	0.45400(−04)	0.51476(−04)	0.45476(−04)	0.45418(−04)	0.45400(−04)	0.10057(−04)	0.38703(−04)
0.60000	0.30590(−06)	0.36932(−04)	0.30667(−06)	0.30609(−06)	0.30590(−06)	0.31892(−07)	0.24078(−06)
0.80000	0.20612(−08)	0.26497(−08)	0.20680(−08)	0.20628(−08)	0.20612(−08)	0.10113(−09)	0.14980(−08)
1.00000	0.13888(−10)	0.19011(−10)	0.13946(−10)	0.13902(−10)	0.13888(−10)	0.32072(−12)	0.93191(−11)

and

$$e_{i+1} = z_{i+1} - u_{i+1} \simeq u^*_{i+1} - u_{i+1}$$

For Runge-Kutta formulas, using the one-step, two half-steps procedure can be very expensive since the cost of computation increases with the number of function evaluations. The following table shows the number of function evaluations per step for pth-order accurate formulas using two half-steps to calculate u^*_{i+1}:

p	2	3	4	5
Evaluations of f per step	5	8	11	14

Take for example the Runge-Kutta-Gill method. The Gill formula requires four function evaluations for the computation of u_{i+1} and seven for u^*_{i+1}. A better procedure is Fehlberg's method (see [5]), which uses a Runge-Kutta formula of higher-order accuracy than used for u_{i+1} to compute u^*_{i+1}. The Runge-Kutta-Fehlberg fourth-order pair of formulas is

$$u_{i+1} = u_i + [\tfrac{25}{216}k_1 + \tfrac{1408}{2565}k_3 + \tfrac{2197}{4104}k_4 - \tfrac{1}{5}k_5], \qquad e_{i+1} = 0(h^5)$$

(1.36)

$$u^*_{i+1} = u_i + [\tfrac{16}{135}k_1 + \tfrac{6656}{12825}k_3 + \tfrac{28561}{56430}k_4 - \tfrac{9}{50}k_5 + \tfrac{2}{55}k_6], \qquad e_{i+1} = 0(h^6),$$

where

$$\begin{aligned}
k_1 &= hf(x_i, u_i)\\
k_2 &= hf(x_i + \tfrac{1}{4}h, u_i + \tfrac{1}{4}k_1)\\
k_3 &= hf(x_i + \tfrac{3}{8}h, u_i + \tfrac{3}{32}k_1 + \tfrac{9}{32}k_2)\\
k_4 &= hf(x_i + \tfrac{12}{13}h, u_i + \tfrac{1932}{2197}k_1 - \tfrac{7200}{2197}k_2 + \tfrac{7296}{2197}k_3)\\
k_5 &= hf(x_i + h, u_i + \tfrac{439}{216}k_1 - 8k_2 + \tfrac{3680}{513}k_3 - \tfrac{845}{4104}k_4)\\
k_6 &= hf(x_i + \tfrac{1}{2}h, u_i - \tfrac{8}{27}k_1 + 2k_2 - \tfrac{3544}{2565}k_3 + \tfrac{1859}{4104}k_4 - \tfrac{11}{40}k_5)
\end{aligned}$$

On first inspection the system (1.36) appears quite complicated, but it can be programmed in a very straightforward way. Notice that the formula for u_{i+1} is fourth-order accurate but requires five function evaluations as compared with the four of the Runge-Kutta-Gill method, which is of the same order accuracy. However, if e_{i+1} is to be estimated, the half-step method using the Runge-Kutta-Gill method requires eleven function evaluations while Eq. (1.36) requires only six—a considerable decrease! The key is to use a pair of formulas with a common set of k_i's. Therefore, if (1.36) is used, as opposed to (1.35), the accuracy is maintained at fourth-order, the stability criteria remains the same, but the cost of computation is significantly decreased. That is why a number of commercially available computer programs (see section on Mathematical Software) use Runge-Kutta-Fehlberg algorithms for solving IVPs.

In this section we have presented methods that increase the order of accuracy, but their stability limitations remain severe. In the next section we discuss methods that have improved stability criteria.

IMPLICIT METHODS

If we once again consider Eq. (1.7) and expand $y(x)$ about the point x_{i+1} using Taylor's theorem with remainder:

$$y(x_i) = y(x_{i+1}) - hy'(x_{i+1}) + \frac{h^2}{2!}y''(\xi_i), \qquad x_i \leqslant \xi_i \leqslant x_{i+1} \tag{1.37}$$

Substitution of (1.7) into (1.37) gives

$$y(x_i) = y(x_{i+1}) - hf(x_{i+1}, y(x_{i+1})) + \frac{h^2}{2!}f_i'(\bar{\xi}, y(\bar{\xi})), \qquad x_i \leqslant \bar{\xi} \leqslant x_{i+1} \tag{1.38}$$

A numerical procedure of (1.7) can be obtained from (1.38) by truncating after the second term:

$$\begin{aligned} u_{i+1} &= u_i + hf_{i+1}, \qquad i = 0, 1, \ldots, N-1, \\ u_0 &= y_0 \end{aligned} \tag{1.39}$$

Equation (1.39) is called the implicit Euler method because the function f is evaluated at the right-hand side of the subinterval. Since the value of u_{i+1} is unknown, (1.39) is nonlinear if f is nonlinear. In this case, one can use a Newton iteration (see Appendix B). This takes the form

$$u_{i+1}^{[s+1]} = h\left[f\bigg|_{u^{[s]}_{i+1}} + \frac{\partial f}{\partial y}\bigg|_{u_{i+1}^{[s]}}(u_{i+1}^{[s+1]} - u_{i+1}^{[s]})\right] + u_i \tag{1.40}$$

or after rearrangement

$$\left(1 - h\frac{\partial f}{\partial y}\right)\bigg|_{u_{i+1}^{[s]}}(u_{i+1}^{[s+1]} - u_{i+1}^{[s]}) = hf\bigg|_{u_{i+1}^{[s]}} + u_i - u_{i+1}^{[s]} \tag{1.41}$$

where $u_{i+1}^{[s]}$ is the sth iterate of u_{i+1}. Iterate on (1.41) until

$$|u_{i+1}^{[s+1]} - u_{i+1}^{[s]}| \leqslant \text{TOL} \tag{1.42}$$

where TOL is a specified absolute error tolerance.

One might ask what has been gained by the implicit nature of (1.39) since it requires more work than, say, the Euler method for solution. If we apply the implicit Euler scheme to (1.17) (λ real),

$$u_{i+1} = u_i + h\lambda u_{i+1}$$

or

$$u_{i+1} = \left(\frac{1}{1 - h\lambda}\right) u_i = \left(\frac{1}{1 - h\lambda}\right)^{i+1} y_0 \tag{1.43}$$

If $\lambda < 0$, then (1.39) is stable for all $h > 0$ or it is unconditionally stable, and never oscillates.

The implicit nature of the method has stabilized the algorithm, but unfortunately the scheme is only first-order accurate. To obtain a higher order of accuracy, combine (1.38) and (1.10) to give

$$2[y(x_{i+1}) - y(x_i)] = h[f_{i+1} + f_i] + 0(h^3) \tag{1.44}$$

The algorithm associated with (1.44) is

$$\begin{aligned} u_{i+1} &= u_i + \frac{h}{2}[f_{i+1} + f_i], \qquad i = 0, 1, \ldots, N - 1, \\ u_0 &= y_0 \end{aligned} \tag{1.45}$$

which is commonly called the trapezoidal rule. Equation (1.45) is second-order accurate, and the stability of the scheme can be examined by applying the method to (1.17), giving (λ real)

$$u_{i+1} = \left[\frac{\left(1 + \dfrac{\lambda h}{2}\right)}{\left(1 - \dfrac{\lambda h}{2}\right)}\right]^{i+1} y_0 \tag{1.46}$$

If $\lambda < 0$, then (1.45) is unconditionally stable, but notice that if $h\lambda < -2$ the method will produce oscillations in the sign of the error. A summary of the stability regions (λ real) for the methods discussed so far is shown in Table 1.5.

From Table 1.5 we see that the Euler method requires a small step-size for stability. Although the criteria for the Runge-Kutta methods are not as

TABLE 1.5 Comparison of Methods Based upon $\frac{dy}{dx} = -\tau y$, $y(0) = 1$, $\tau > 0$ and is a real constant

Method	Stable Step-Size, No Oscillations	Stable Step-Size, Oscillations	Unstable Step-Size	Order of Accuracy
Euler (1.11)	$h\tau < 1$	$1 \leq h\tau \leq 2$	$2 < h\tau$	1
Second-order Runge-Kutta (1.33)	$h\tau \leq 2$	None	$2 < h\tau$	2
Runge-Kutta-Gill (1.35)	$h\tau \leq 2.8$	None	$2.8 < h\tau$	4
Implicit Euler (1.39)	$h\tau < \infty$	None	None	1
Trapezoidal (1.45)	$h\tau < 2$	$2 \leq h\tau \leq \infty$	None	2

stringent as for the Euler method, stable step-sizes for these schemes are also quite small. The trapezoidal rule requires a small step-size to avoid oscillations but is stable for any step-size, while the implicit Euler method is always stable. The previous two algorithms require more arithmetic operations than the Euler or Runge-Kutta methods when f is nonlinear due to the Newton iteration, but are typically used for solution of certain types of problems (see section on stiffness).

In Table 1.5 we once again see the dilemma of stability versus accuracy. In the following section we outline one technique for increasing the accuracy when using any method.

EXTRAPOLATION

Suppose we solve a problem with a step-size of h giving the solution u_i at x_i, and also with a step-size $h/2$ giving the solution ω_i at x_i. If an Euler method is used to obtain u_i and ω_i, then the error is proportional to the step-size (first-order accurate). If $y(x_i)$ is the exact solution at x_i, then

$$u_i \simeq y(x_i) + \phi h \qquad \textbf{(1.47)}$$

$$\omega_i \simeq y(x_i) + \phi \frac{h}{2}$$

where ϕ is a constant. Eliminating ϕ from (1.47) gives

$$y(x_i) = 2\omega_i - u_i \qquad \textbf{(1.48)}$$

If the error formulas (1.47) are exact, then this procedure gives the exact solution. Since the formulas (1.47) usually only apply as $h \to 0$, then (1.48) is only an approximation, but it is expected to be a more accurate estimate than either ω_i or u_i. The same procedure can be used for higher-order methods. For the trapezoidal rulc

$$u_i \simeq y(x_i) + \phi h^2$$

$$\omega_i \simeq y(x_i) + \phi \left(\frac{h}{2}\right)^2 \qquad \textbf{(1.49)}$$

$$y(x_i) \simeq \frac{4\omega_i - u_i}{3}$$

EXAMPLE 4

The batch still shown in Figure 1.3 initially contains 25 moles of n-octane and 75 moles of n-heptane. If the still is operated at a constant pressure of 1 atmosphere (atm), compute the final mole fraction of n-heptane, x_H^f, if the remaining solution in the still, S^f, totals 10 moles.

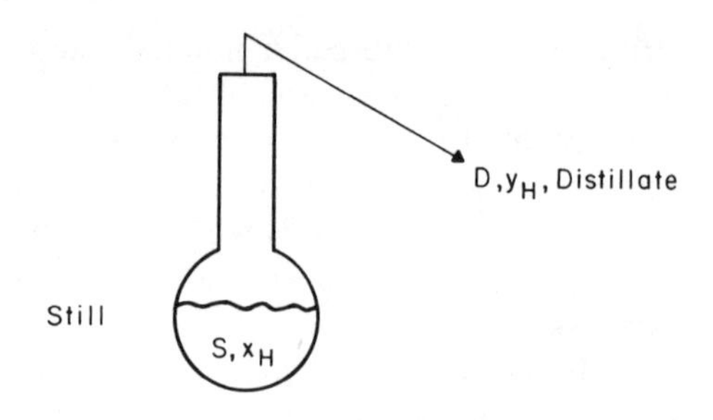

FIGURE 1.3 Batch still.

Data: At 1 atm total pressure, the relationship between x_H and the mole fraction of n-heptane in the vapor phase, y_H, is

$$y_H = \frac{2.16x_H}{1 + 1.16\,x_H}$$

SOLUTION

An overall material balance is

$$dS = -dD$$

A material balance of n-heptane gives

$$d(x_H S) = -y_H dD$$

Combination of these balances yields

$$\int_{S_o}^{S^f} \frac{dS}{S} = \int_{x_H^0}^{x_H^f} \frac{dx_H}{y_H - x_H}$$

where $S_0 = 100$, $S^f = 10$, $x_H^0 = 0.75$.
Substitute for y_H and integrate to give

$$\left(\frac{S^f}{S_0}\right) = \left(\frac{1 - x_H^0}{1 - x_H^f}\right)\left[\left(\frac{1 - x_H^0}{1 - x_H^f}\right)\left(\frac{x_H^f}{x_H^0}\right)\right]^{1/1.16}$$

and

$$x_H^f = 0.37521825$$

Physically, one would expect x_H to decrease with time since heptane is lighter than octane and would flash in greater amounts than would octane. Now compare the analytical solution to the following numerical solutions. First, reformulate the differential equation by defining

$$t = \frac{S_0 - S}{S_0 - S^f}$$

so that

$$0 \leq t \leq 1$$

Thus:

$$\frac{dx_H}{dt} = 1.16 \frac{(S^f - S_0)}{(S_0(1 - t) + S^f t)} \frac{x_H(1 - x_H)}{(1 + 1.16x_H)}, \qquad x_H = x_H^0 \quad \text{at} \quad t = 0$$

If an Euler method is used, the results are shown in Table 1.6. From a practical standpoint, all the values in Table 1.6 would probably be sufficiently accurate for design purposes, but we provide the large number of significant figures to illustrate the extrapolation method. A simple Euler method is first-order accurate, and so the truncation error should be proportional to $h(1/N)$. This is shown in Table 1.6. Also notice that the error in the extrapolated Euler method decreases faster than that in the Euler method with increasing N. The truncation error of the extrapolation is approximately the square of the error in the basic method. In this example one can see that improved accuracy with less computation is achieved by extrapolation. Unfortunately, the extrapolation is successful only if the step-size is small enough for the truncation error formula to be reasonably accurate. Some nonlinear problems require extremely small step-sizes and can be computationally unreasonable.

Extrapolation is one method of increasing the accuracy, but it does not change the stability of a method. There are commercial packages that employ extrapolation (see section on Mathematical Software), but they are usually based upon Runge-Kutta methods instead of the Euler or trapezoidal rule as outlined

TABLE 1.6 Errors in the Euler Method and the Extrapolated Euler Method for Example 4

Number of Steps	Total Number of Steps	Absolute Value of the Error
Euler Method		
50	50	0.01373
100	100	0.00675
200	200	0.00335
400	400	0.00166
800	800	0.00083
1,600	1,600	0.00041
Extrapolated Euler Method		
50–100	150	0.000220
100–200	300	0.000056
200–400	600	0.000013
400–800	1,200	0.000003
800–1600	2,400	0.000001

above. In the following section we describe techniques currently being used in software packages for which stability, accuracy, and computational efficiency have been addressed in detail (see, for example, [5]).

MULTISTEP METHODS

Multistep methods make use of information about the solution and its derivative at more than one point in order to extrapolate to the next point. One specific class of multistep methods is based on the principle of numerical integration. If the differential equation $y' = f(x, y)$ is integrated from x_i to x_{i+1}, we obtain

$$\int_{x_i}^{x_{i+1}} y'\, dx = \int_{x_i}^{x_{i+1}} f(x, y(x))\, dx$$

or

$$y(x_{i+1}) = y(x_i) + \int_{x_i}^{x_{i+1}} f(x, y(x))\, dx \tag{1.50}$$

To carry out the integration in (1.50), approximate $f(x, y(x))$ by a polynomial that interpolates $f(x, y(x))$ at k points, $x_i, x_{i-1}, \ldots, x_{i-k+1}$. If the Newton backward formula of degree $k-1$ is used to interpolate $f(x, y(x))$, then the Adams-Bashforth formulas [1] are generated and are of the form

$$u_{i+1} = u_i + h \sum_{j=1}^{k} b_j u'_{i-j+1} \tag{1.51}$$

where

$$u'_j = f(x_j, u_j)$$

This is called a k-step formula because it uses information from the previous k steps. Note that the Euler formula is a one-step formula ($k = 1$) with $b_1 = 1$. Alternatively, if one begins with (1.51), the coefficients b_j can be chosen by assuming that the past values of u are exact and equating like powers of h in the expansion of (1.51) and of the local solution z_{i+1} about x_i. In the case of a three-step formula

$$u_{i+1} = u_i + h[b_1 u'_i + b_2 u'_{i-1} + b_3 u'_{i-2}]$$

Substituting values of z into this and expanding about x_i gives

$$z_{i+1} = z_i + hz'_i[b_1 + b_2 + b_3] - h^2 z''_i[b_2 + 2b_3] + \frac{h^3}{2!} z'''_i[b_2 + 4b_3] + \ldots$$

where

$$z'_{i-1} = z'_i - hz''_i + \frac{h^2}{2!} z'''_i + \ldots$$

$$z'_{i-2} = z'_i - 2hz''_i + \frac{4h^2}{2!} z'''_i + \ldots$$

The Taylor's series expansion of z_{i+1} is

$$z_{i+1} = z_i + hz_i' + \frac{h^2}{2!} z_i'' + \frac{h^3}{3!} z_i''' + \dots$$

and upon equating like power of h, we have

$$b_1 + b_2 + b_3 = 1$$

$$b_2 + 2b_3 = -\tfrac{1}{2}$$

$$b_2 + 4b_3 = \tfrac{1}{3}$$

The solution of this set of linear equations is $b_1 = \frac{23}{12}$, $b_2 = -\frac{16}{12}$, and $b_3 = \frac{5}{12}$. Therefore, the three-step Adams-Bashforth formula is

$$u_{i+1} = u_i + \frac{h}{12} [23u_i' - 16u_{i-1}' + 5u_{i-2}'] \tag{1.52}$$

with an error $e_{i+1} = 0(h^4)$ [generally $e_{i+1} = 0(h^{k+1})$ for any value of k; for example, in (1.52) $k = 3$].

A difficulty with multistep methods is that they are not self-starting. In (1.52) values for u_i, u_i', u_{i-1}', and u_{i-2}' are needed to compute u_{i+1}. The traditional technique for computing starting values has been to use Runge-Kutta formulas of the same accuracy since they only require u_0 to get started. An alternative procedure, which turns out to be more efficient, is to use a sequence of s-step formulas with $s = 1, 2, \dots, k$ [6]. The computation is started with the one-step formulas in order to provide starting values for the two-step formula and so on. Also, the problem of getting started arises whenever the step-size h is changed. This problem is overcome by using a k-step formula whose coefficients (the b_i's) depend upon the past step-sizes ($h_s = x_s - x_{s-1}$, $s = i, i - 1, \dots, i - k + 1$) (see [6]). This kind of procedure is currently used in commercial multistep routines.

The previous multistep methods can be derived using polynomials that interpolated at the point x_i and at points backward from x_i. These are sometimes known as formulas of explicit type. Formulas of implicit type can also be derived by basing the interpolating polynomial on the point x_{i+1}, as well as on x_i and points backward from x_i. The simplest formula of this type is obtained if the integral is approximated by the trapezoidal formula. This leads to

$$u_{i+1} = u_i + \frac{h}{2} [f(x_i, u_i) + f(x_{i+1}, u_{i+1})]$$

which is Eq. (1.45). If f is nonlinear, u_{i+1} cannot be solved for directly. However, we can attempt to obtain u_{i+1} by means of iteration. Predict a first approximation $u_{i+1}^{[0]}$ to u_{i+1} by using the Euler method

$$u_{i+1}^{[0]} = u_i + hf_i \tag{1.53}$$

Then compute a corrected value with the trapezoidal formula

$$u_{i+1}^{[s+1]} = u_i + \frac{h}{2}[f_i + f(u_{i+1}^{[s]})], \qquad s = 0, 1, \ldots \tag{1.54}$$

For most problems occurring in practice, convergence generally occurs within one or two iterations. Equations (1.53) and (1.54) used as outlined above define the simplest predictor-corrector method.

Predictor-corrector methods of higher-order accuracy can be obtained by using the multistep formulas such as (1.52) to predict and by using corrector formulas of type

$$u_{i+1} = u_i + h \sum_{j=0}^{k} b_j u'_{i-j+1} \tag{1.55}$$

Notice that j now sums from zero to k. This class of corrector formulas is called the Adams-Moulton correctors. The b_j's of the above equation can be found in a manner similar to those in (1.52). In the case of $k = 2$,

$$u_{i+1} = u_i + \frac{h}{12}[5u'_{i+1} + 8u'_i - u'_{i-1}] \tag{1.56}$$

with a local truncation error of $0(h^4)$. A similar procedure to that outlined for the use of (1.53) and (1.54) is constructed using (1.52) as the predictor and (1.56) as the corrector. The combination (1.52), (1.56) is called the Adams-Moulton predictor-corrector pair of formulas.

Notice that the error in each of the formulas (1.52) and (1.56) is $0(h^4)$. Therefore, if e_{i+1} is to be estimated, the difference

$$|u_{i+1}^* - u_{i+1}|, \qquad u_{i+1}^* \text{ from (1.56)}, \quad u_{i+1} \text{ from (1.52)}$$

would be a poor approximation. More precise expressions for the errors in these formulas are [5]

$$e_{i+1} = \tfrac{3}{8}h^4 y''''(\xi), \qquad \text{for} \quad (1.52)$$

$$e_{i+1}^* = -\tfrac{1}{24}h^4 y''''(\xi^*), \qquad \text{for} \quad (1.56)$$

where $x_{i-2} < \xi$ and $\xi^* < x_{i+1}$. Assume that $\xi^* = \xi$ (this would be a good approximation for small h), then subtract the two expressions.

$$e_{i+1}^* - e_{i+1} = u_{i+1}^* - u_{i+1} = -\tfrac{5}{12}h^4 y''''(\xi)$$

Solving for $h^4 y''''(\xi)$ and substituting into the expression e_{i+1}^* gives

$$|e_{i+1}^*| \simeq \tfrac{1}{10}|u_{i+1}^* - u_{i+1}|$$

Since we had to make a simplifying assumption to obtain this result, it is better to use a more conservative coefficient, say $\frac{1}{8}$. Hence,

$$|e_{i+1}^*| \simeq \tfrac{1}{8}|u_{i+1}^* - u_{i+1}| \tag{1.57}$$

Note that this is an error estimate for the more accurate value so that u_{i+1}^* can be used as the numerical solution rather than u_{i+1}. This type of analysis is not used in the case of Runge-Kutta formulas because the error expressions are very complicated and difficult to manipulate in the above fashion.

Since the Adams-Bashforth method [Eq. (1.51)] is explicit, it possesses poor stability properties. The region of stability for the implicit Adams-Moulton method [Eq. (1.55)] is larger by approximately a factor of 10 than the explicit Adams-Bashforth method, although in both cases the region of stability decreases as k increases (see p. 130 of [4]). For the Adams-Moulton predictor-corrector pair, the exact regions of stability are not well defined, but the stability limitations are less severe than for explicit methods and depend upon the number of corrector iterations [4].

The multistep integration formulas listed above can be represented by the generalized equation:

$$u_{i+1} = \sum_{j=1}^{k_1} a_{i+1,j}\, u_{i-j+1} + h_{i+1} \sum_{j=0}^{k_2} b_{i+1,j}\, u'_{i-j+1} \qquad \textbf{(1.58)}$$

which allows for variable step-size through h_{i+1}, $a_{i+1,j}$, and $b_{i+1,j}$. For example, if $k_1 = 1$, $a_{i+1,1} = 1$ for all i, $b_{i+1,j} = b_{i,j}$ for all i, and $k_2 = q - 1$, then a qth-order implicit formula is obtained. Further, if $b_{i+1,0} = 0$, then an explicit formula is generated. Computationally these methods are very efficient. If an explicit formula is used, only a single function evaluation is needed per step. Because of their poor stability properties, explicit multistep methods are rarely used in practice. The use of predictor-corrector formulas does not necessitate the solution of nonlinear equations and requires $S + 1$ (S is the number of corrector iterations) function evaluations per step in x. Since S is usually small, fewer function evaluations are required than from an equivalent order of accuracy Runge-Kutta method and better stability properties are achieved. If a problem requires a large stability region (see section of stiffness), then implicit backward formulas must be used. If (1.58) represents an implicit backward formula, then it is given by

$$u_{i+1} = \sum_{j=1}^{k_1} a_{i+1,j}\, u_{i-j+1} + h_{i+1}\, b_{i+1,0}\, u'_{i+1}$$

or

$$u_{i+1} = b_{i+1,0}\, h_{i+1}\, f(u_{i+1}) + \phi_i \qquad \textbf{(1.59)}$$

where ϕ_i is the grouping of all known information. If a Newton iteration is performed on (1.59), then

$$\left[1 - b_{i+1,0}\, h_{i+1} \left.\frac{\partial f}{\partial y}\right|_{u_{i+1}^{[s]}}\right] \left[u_{i+1}^{[s+1]} - u_{i+1}^{[s]}\right]$$

$$= b_{i+1,0}\, h_{i+1} \left. f \right|_{u_{i+1}^{[s]}} + \phi_i - u_{i+1}^{[s]}, \qquad s = 0, 1, \ldots \qquad \textbf{(1.60)}$$

Therefore, the derivative $\partial f/\partial y$ must be calculated and the function f evaluated at each iteration. One must "pay" in computation time for the increased stability. The order of accuracy of implicit backward formulas is determined by the value of k_1. As k_1 is increased, higher accuracy is achieved, but at the expense of decreased stability (see Chapter 11 of [4]).

Multistep methods are frequently used in commercial routines because of their combined accuracy, stability, and computational efficiency properties (see section on Mathematical Software). Other high-order methods for handling problems that require large regions of stability are discussed in the following section.

HIGH-ORDER METHODS BASED ON KNOWLEDGE OF $\partial f/\partial y$

A variety of methods that make use of $\partial f/\partial y$ has been proposed to solve problems that require large stability regions. Rosenbrock [7] proposed an extension of the explicit Runge-Kutta process that involved the use of $\partial f/\partial y$. Briefly, if one allows the summation in (1.25) to go from 1 to j, i.e., an implicit Runge-Kutta method, then,

$$\overline{k}_j = hf\left(u_i + \sum_{l=1}^{j} a_{jl}\overline{k}_l\right) \tag{1.61}$$

If $\overline{k}_j$ is expanded,

$$\overline{k}_j = hf\left(u_i + \sum_{l=1}^{j-1} a_{jl}\overline{k}_l\right) + h\frac{\partial f}{\partial y}\left(u_i + \sum_{l=1}^{j-1} a_{jl}\overline{k}_l\right) a_{jj}\overline{k}_j \tag{1.62}$$

and rearranged to give

$$\left[1 - ha_{jj}\frac{\partial f}{\partial y}\left(u_i + \sum_{l=1}^{j-1} a_{jl}\overline{k}_l\right)\right]\overline{k}_j = hf\left(u_i + \sum_{l=1}^{j-1} a_{jl}\overline{k}_l\right) \tag{1.63}$$

the method is called a semi-implicit Runge-Kutta method. In the function f, it is assumed that the independent variable x does not appear explicitly, i.e., it is autonomous. Equation (1.63) is used with

$$u_{i+1} = u_i + \sum_{j=1}^{\nu} \omega_j\overline{k}_j \tag{1.64}$$

to specify the method. Notice that if the bracketed term in (1.63) is replaced by 1, then (1.63) is an explicit Runge-Kutta formula. Calahan [8], Allen [9], and Caillaud and Padmanabhan [10] have developed these methods into algorithms and have shown that they are unconditionally stable with no oscillations in the solution. Stabilization of these algorithms is due to the bracketed term in (1.63). We will return to this semi-implicit method in the section Mathematical Software.

Other methods that are high-order, are stable, and do not oscillate are the second- and third-order semi-implicit methods of Norsett [11], more recently the diagonally implicit methods of Alexander [12], and those of Bui and Bui [13] and Burka [14].

STIFFNESS

Up to this point we have limited our discussion to a single differential equation. Before looking at systems of differential equations, an important characteristic of systems, called stiffness, is illustrated.

Suppose we wish to model the reaction path $A \underset{k_2}{\overset{k_1}{\rightleftharpoons}} B$ starting with pure A. The reaction path can be described by

$$\frac{dC_A}{dt} = -k_1 C_A + k_2 C_B \tag{1.65}$$

where

$$C_A = C_A^0 \text{ at } t = 0$$

$$C_A = \text{concentration of A}$$

$$t = \text{time}$$

One can define $y_1 = (C_A - C_A^{eq})/(C_A^0 - C_A^{eq})$ where C_A^{eq} is the equilibrium value of C_A $(t \rightarrow \infty)$. Equation (1.65) becomes

$$\frac{dy_1}{dt} = -(k_1 + k_2)\, y_1, \qquad y_1 = 1 \quad \text{at} \quad t = 0 \tag{1.66}$$

If $k_1 = 1000$ and $k_2 = 1$, then the solution of (1.66) is

$$y_1 = e^{-1001t} \tag{1.67}$$

If one uses the Euler method to solve (1.66), then

$$h < \tfrac{1}{1001}$$

for stability. The time required to observe the full evolution of the solution is very short. If one now wishes to follow the reaction path $B \overset{k_3}{\rightarrow} D$, then

$$\frac{dC_B}{dt} = -k_3 C_B, \qquad C_B = C_B^0 \quad \text{at} \quad t = 0 \tag{1.68}$$

If $k_3 = 1$ and $y_2 = C_B/C_B^0$, then the solution of (1.68) is

$$y_2 = e^{-t} \tag{1.69}$$

If the Euler method is applied to (1.68), then

$$h < 1$$

for stability. The time required to observe the full evolution of the solution is long when compared with that required by (1.66). Next suppose we wish to simulate the reaction pathway

$$\mathrm{A} \underset{k_2}{\overset{k_1}{\rightleftharpoons}} \mathrm{B} \overset{k_3}{\rightarrow} \mathrm{D}$$

The governing differential equations are

$$\frac{dC_A}{dt} = -k_1 C_A + k_2 C_B$$

$$\frac{dC_B}{dt} = k_1 C_A - (k_2 + k_3) C_B, \tag{1.70}$$

$$C_A = C_A^0,\ C_B = 0 \quad \text{at} \quad t = 0$$

This system can be written as

$$\frac{d\mathbf{y}}{dt} = Q\mathbf{y} = \mathbf{f},\ \mathbf{y}(0) = [1, 0]^T \tag{1.71}$$

where

T designates the transpose

$$\mathbf{y} = \left[\frac{C_A}{C_A^0}\ \ \frac{C_B}{C_A^0}\right]^T$$

$$Q = \begin{bmatrix} -k_1 & k_2 \\ k_1 & -(k_2 + k_3) \end{bmatrix}$$

$$\mathbf{f} = [f_1, f_2]^T$$

The solution of (1.71) is

$$y_1 = \tfrac{1000}{1001} e^{-1001t} + \tfrac{1}{1001} e^{-t}$$

$$y_2 = -\tfrac{1000}{1001} e^{-1001t} + \tfrac{1000}{1001} e^{-t} \tag{1.72}$$

A plot of (1.72) is shown in Figure 1.4. Notice that y_1 decays very rapidly, as would (1.67), whereas y_2 requires a long time to trace its full evolution, as would (1.69). If (1.71) is solved by the Euler method

$$h < \frac{1}{|\lambda_{\mathscr{E}}|_{max}} \tag{1.73}$$

where $|\lambda_{\mathscr{E}}|_{max}$ is the absolute value of the largest eigenvalue of Q. We have the unfortunate situation with systems of equations that the largest step-size is governed by the largest eigenvalue while the integration time for full evolution of the solution is governed by the smallest eigenvalue (slowest decay rate). This property of systems is called stiffness and can be quantified by the stiffness ratio

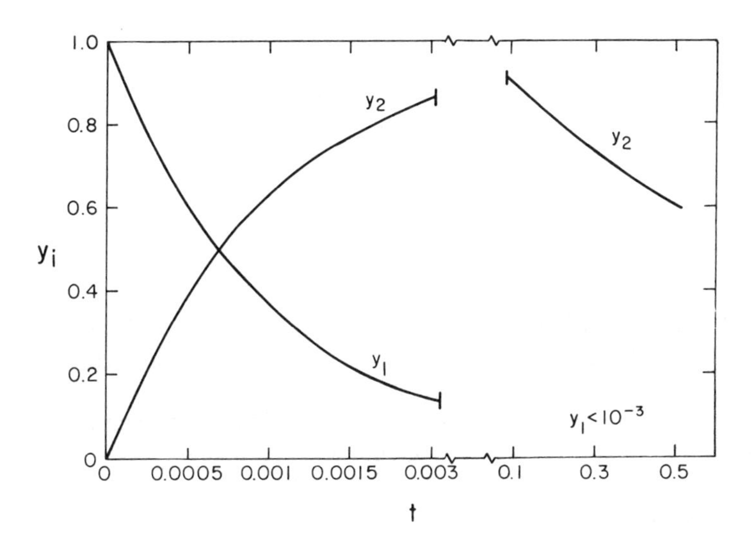

FIGURE 1.4 Results from Eq. (1.72).

[15] SR,

$$\mathrm{SR} = \frac{\max_i |\text{real part of } \lambda_{\mathscr{E}_i}|}{\min_i |\text{real part of } \lambda_{\mathscr{E}_i}|}, \qquad \text{real part of } \lambda_{\mathscr{E}_i} < 0, \qquad i = 1, \ldots, m, \tag{1.74}$$

$$m = \text{number of equations in the system}$$

which allows for imaginary eigenvalues. Typically SR = 20 is not stiff, SR = 10^3 is stiff, and SR = 10^6 is very stiff. From (1.72) SR = $\frac{1001}{1} \simeq 10^3$, and the system (1.71) is stiff. If the system of equations (1.71) were nonlinear, then a linearization of (1.71) gives

$$\frac{d\mathbf{y}}{dt} = \boldsymbol{Q}(t_i)\mathbf{y}(t_i) + \boldsymbol{J}(t_i)(\mathbf{y} - \mathbf{y}(t_i)) \tag{1.75}$$

where

$$\begin{aligned} \mathbf{y}(t_i) &= \text{vector } \mathbf{y} \text{ evaluated at time } t_i \\ \boldsymbol{Q}(t_i) &= \text{matrix } \boldsymbol{Q} \text{ evaluated at time } t_i \\ \boldsymbol{J}(t_i) &= \text{matrix } \boldsymbol{J} \text{ evaluated at time } t_i \end{aligned}$$

$$\boldsymbol{J} = \begin{bmatrix} \dfrac{\partial f_1}{\partial y_1} & \dfrac{\partial f_1}{\partial y_2} \\ \dfrac{\partial f_2}{\partial y_1} & \dfrac{\partial f_2}{\partial y_2} \end{bmatrix}$$

The matrix J is called the Jacobian matrix, and in general is

$$J = \begin{bmatrix} \frac{\partial f_1}{\partial y_1}, \frac{\partial f_1}{\partial y_2}, \ldots, \frac{\partial f_1}{\partial y_m} \\ \vdots \\ \frac{\partial f_m}{\partial y_1}, \frac{\partial f_m}{\partial y_2}, \ldots, \frac{\partial f_m}{\partial y_m} \end{bmatrix}$$

For nonlinear problems the stiffness is based upon the eigenvalues of J and thus applies only to a specific time, and it may change with time. This characteristic of systems makes a problem both interesting and difficult. We need to classify the stiffness of a given problem in order to apply techniques that "perform" well for that given magnitude of stiffness. Generally, implicit methods "outperform" explicit methods on stiff problems because of their less rigid stability criterion. Explicit methods are best suited for nonstiff equations.

SYSTEMS OF DIFFERENTIAL EQUATIONS

A straightforward extension of (1.11) to a system of equations is

$$\begin{aligned} \mathbf{u}_{i+1} &= \mathbf{u}_i + h\mathbf{f}(x_i, \mathbf{u}_i), \qquad i = 0, 1, \ldots, N-1 \\ \mathbf{u}_0 &= \mathbf{y}_0 \end{aligned} \tag{1.76}$$

Likewise, the implicit Euler becomes

$$\begin{aligned} \mathbf{u}_{i+1} &= \mathbf{u}_i + h\mathbf{f}(x_i, \mathbf{u}_{i+1}), \qquad i = 0, 1, \ldots, N-1 \\ \mathbf{u}_0 &= \mathbf{y}_0 \end{aligned} \tag{1.77}$$

while the trapezoid rule gives

$$\begin{aligned} \mathbf{u}_{i+1} &= \mathbf{u}_i + \frac{h}{2}[\mathbf{f}(x_i, \mathbf{u}_{i+1}) + \mathbf{f}(x_i, \mathbf{u}_i)], \qquad i = 0, 1, \ldots, N-1 \\ \mathbf{u}_0 &= \mathbf{y}_0 \end{aligned} \tag{1.78}$$

For a system of equations the Runge-Kutta-Fehlberg method is

$$\begin{aligned} \mathbf{u}_{i+1} &= \mathbf{u}_i + \left[\tfrac{25}{216}\mathbf{k}_1 + \tfrac{1408}{2565}\mathbf{k}_3 + \tfrac{2197}{4104}\mathbf{k}_4 - \tfrac{1}{5}\mathbf{k}_5\right] \\ \mathbf{u}^*_{i+1} &= \mathbf{u}_i + \left[\tfrac{16}{135}\mathbf{k}_1 + \tfrac{6656}{12825}\mathbf{k}_3 + \tfrac{28561}{56430}\mathbf{k}_4 - \tfrac{9}{50}\mathbf{k}_5 + \tfrac{2}{55}\mathbf{k}_6\right] \end{aligned} \tag{1.79}$$

where

$$\mathbf{k}_i = \left[k_i^{\{1\}}, k_i^{\{2\}}, \ldots, k_i^{\{m\}}\right]^T$$

and, for example,

$$k_1^{\{j\}} = hf_j(x_i, u_i^{\{1\}}, u_i^{\{2\}}, \ldots, u_i^{\{m\}}), \qquad j = 1, \ldots, m$$

$$\phi_{ji} = u_i^{\{j\}} + \tfrac{1}{4}k_1^{\{j\}}$$

$$k_2^{\{j\}} = hf_j(x_i + \tfrac{1}{4}h, \phi_{1i}, \phi_{2i}, \ldots, \phi_{mi}), \qquad j = 1, \ldots, m$$

$$\vdots$$

The Adams-Moulton predictor-corrector formulas for a system of equations are

$$\begin{aligned} \mathbf{u}_{i+1} &= \mathbf{u}_i + \frac{h}{12}[23\mathbf{u}_i' - 16\mathbf{u}_{i-1}' + 5\mathbf{u}_{i-2}'] \\ \mathbf{u}_{i+1}^* &= \mathbf{u}_i + \frac{h}{12}[5\mathbf{u}_{i+1}' + 8\mathbf{u}_i' - \mathbf{u}_{i-1}'] \end{aligned} \tag{1.80}$$

An algorithm using the higher-order method of Caillaud and Padmanabhan [10] was formulated by Michelsen [16] by choosing the parameters in (1.63) so that the same factor multiplies each $\bar{\mathbf{k}}_i$, thus minimizing the work involved in matrix inversion. The final scheme is

$$\begin{aligned} \mathbf{u}_{i+1} &= \mathbf{u}_i + \omega_1\bar{\mathbf{k}}_1 + \omega_2\bar{\mathbf{k}}_1 + \omega_3\bar{\mathbf{k}}_3, \qquad i = 0, 1, \ldots, N-1 \\ \mathbf{u}_0 &= \mathbf{y}_0 \\ \bar{\mathbf{k}}_1 &= h\left[\boldsymbol{I} - ha_1\frac{\partial \mathbf{f}}{\partial \mathbf{y}}(\mathbf{u}_i)\right]^{-1}\mathbf{f}(\mathbf{u}_i) \\ \bar{\mathbf{k}}_2 &= h\left[\boldsymbol{I} - ha_1\frac{\partial \mathbf{f}}{\partial \mathbf{y}}(\mathbf{u}_i)\right]^{-1}\mathbf{f}(\mathbf{u}_i + b_2\bar{\mathbf{k}}_1) \\ \bar{\mathbf{k}}_3 &= h\left[\boldsymbol{I} - ha_1\frac{\partial \mathbf{f}}{\partial \mathbf{y}}(\mathbf{u}_i)\right]^{-1}(b_{31}\bar{\mathbf{k}}_1 + b_{32}\bar{\mathbf{k}}_2) \end{aligned} \tag{1.81}$$

where $\boldsymbol{I}$ is the identity matrix,

$$\begin{aligned} a_1 &= 0.43586659 \\ b_2 &= 0.75 \\ b_{31} &= \frac{-1}{6a_1}(8a_1^2 - 2a_1 + 1) \\ b_{32} &= \frac{2}{9a_1}(6a_1^2 - 6a_1 + 1) \\ \omega_1 &= \tfrac{11}{27} - b_{31} \\ \omega_2 &= \tfrac{16}{27} - b_{32} \\ \omega_3 &= 1.0 \end{aligned}$$

As previously stated, the independent variable x must not explicitly appear in $\mathbf{f}$. If x does explicitly appear in $\mathbf{f}$, then one must reformulate the system of equations by introducing a new integration variable, t, and let

$$\frac{dx}{dt} = 1 \tag{1.82}$$

be the $(m + 1)$ equation in the system.

EXAMPLE 5

Referring to Example 1, if we now consider the reactor to be adiabatic instead of isothermal, then an energy balance must accompany the material balance. Formulate the system of governing differential equations and evaluate the stiffness. Write down the Euler and the Runge-Kutta-Fehlberg methods for this problem.

Data

$$C_p = 12.17 \times 10^4 \text{ J/(kmole·°C)}$$

$$-\Delta H_r = 2.09 \times 10^8 \text{ J/kmole}$$

SOLUTION

Let $T^* = T/T°$, $T° = 423$ K (150°C). For the "short" reactor,

$$\frac{dy}{dx} = -0.1744 \exp\left[\frac{3.21}{T^*}\right] y \qquad \text{(material balance)}$$

$$\frac{dT^*}{dx} = 0.06984 \exp\left[\frac{3.21}{T^*}\right] y \qquad \text{(energy balance)}$$

$$y = 1,\ T^* = 1 \quad \text{at} \quad x = 0$$

First, check to see if stiffness is a problem. To do this the transport equations can be linearized and the Jacobian matrix formed.

$$\mathrm{J} = \begin{bmatrix} -0.1744 \exp\left(\dfrac{3.21}{T^*}\right) & \dfrac{0.56}{(T^*)^2} \exp\left(\dfrac{3.21}{T^*}\right) y \\ \\ 0.06984 \exp\left(\dfrac{3.21}{T^*}\right) & \dfrac{-0.224}{(T^*)^2} \exp\left(\dfrac{3.21}{T^*}\right) y \end{bmatrix}$$

At the inlet $T^* = 1$ and $y = 1$, and the eigenvalues of $\boldsymbol{J}$ are approximately $(6.3, -7.6)$. Since T^* should increase as y decreases, for example, if $T^* = 1.12$ and $y = 0.5$, then the eigenvalues of $\boldsymbol{J}$ are approximately $(3.0, -4.9)$. From the stiffness ratio, one can see that this problem is not stiff.

ROOTS OF

$\det(\lambda I - J) = 0$

Euler:

$$u_{i+1}^{\{1\}} = u_i^{\{1\}} - 0.1744 \exp\left[\frac{3.21}{u_i^{\{2\}}}\right] u_i^{\{1\}} h$$

$$u_{i+1}^{\{2\}} = u_i^{\{2\}} + 0.06984 \exp\left[\frac{3.21}{u_i^{\{2\}}}\right] u_i^{\{1\}} h$$

$$u_0^{\{1\}} = 1$$

$$u_0^{\{2\}} = 1$$

Runge-Kutta-Fehlberg:

$$u_{i+1}^{\{1\}} = u_i^{\{1\}} + [C1 \cdot k_1^{\{1\}} + C2 \cdot k_3^{\{1\}} + C3 \cdot k_4^{\{1\}} + C4 \cdot k_5^{\{1\}}]$$

$$u_{i+1}^{\{2\}} = u_i^{\{2\}} + [C1 \cdot k_1^{\{2\}} + C2 \cdot k_3^{\{2\}} + C3 \cdot k_4^{\{2\}} + C4 \cdot k_5^{\{2\}}]$$

$$u_{i+1}^{\{1\}*} = u_i^{\{1\}} + [C5 \cdot k_1^{\{1\}} + C6 \cdot k_3^{\{1\}} + C7 \cdot k_4^{\{1\}} + C8 \cdot k_5^{\{1\}} + C9 \cdot k_6^{\{1\}}]$$

$$u_{i+1}^{\{2\}*} = u_i^{\{2\}} + [C5 \cdot k_1^{\{2\}} + C6 \cdot k_3^{\{2\}} + C7 \cdot k_4^{\{2\}} + C8 \cdot k_5^{\{2\}} + C9 \cdot k_6^{\{2\}}]$$

$$C1 = \tfrac{25}{216}, \qquad C5 = \tfrac{16}{135}$$

$$C2 = \tfrac{1408}{2565}, \qquad C6 = \tfrac{6656}{12825}$$

$$C3 = \tfrac{2197}{4104}, \qquad C7 = \tfrac{28561}{56430}$$

$$C4 = -\tfrac{1}{5}, \qquad C8 = -\tfrac{9}{50}$$

$$C9 = \tfrac{9}{55}$$

Define

$$F1(\mathrm{A}, \mathrm{B}) = -0.1744 \exp\left[\frac{3.21}{\mathrm{B}}\right] \mathrm{A}$$

$$F2(\mathrm{A}, \mathrm{B}) = 0.06984 \exp\left[\frac{3.21}{\mathrm{B}}\right] \mathrm{A}$$

then

$$k_1^{\{1\}} = hF1(u_i^{\{1\}}, u_i^{\{2\}})$$

$$k_1^{\{2\}} = hF2(u_i^{\{1\}}, u_i^{\{2\}})$$

$$k_2^{\{1\}} = hF1(u_i^{\{1\}} + \tfrac{1}{4}k_1^{\{1\}}, u_i^{\{2\}} + \tfrac{1}{4}k_1^{\{2\}})$$

$$k_2^{\{2\}} = hF2(u_i^{\{1\}} + \tfrac{1}{4}k_1^{\{1\}}, u_i^{\{2\}} + \tfrac{1}{4}k_1^{\{2\}})$$

$$\vdots$$

STEP-SIZE STRATEGIES

Thus far we have only concerned ourselves with constant step-sizes. Variable step-sizes can be very useful for (1) controlling the local truncation error and (2) improving efficiency during solution of a stiff problem. This is done in all of the commercial programs, so we will discuss each of these points in further detail.

Gear [4] estimates the local truncation error and compares it with a desired error, TOL. If the local truncation error has been achieved using a step-size h_1,

$$e = \phi h_1^{p+1} \tag{1.83}$$

Since we wish the error to equal TOL,

$$\text{TOL} = \phi h_2^{p+1} \tag{1.84}$$

Combination of (1.83) and (1.84) gives

$$h_2 = h_1 \left[\frac{\text{TOL}}{e} \right]^{1/(p+1)} \tag{1.85}$$

Equation (1.83) is method-dependent, so we will illustrate the procedure with a specific method. If we solve a given problem using the Euler method,

$$u_{i+1} = u_i + h_1 f(u_i) \tag{1.86}$$

and the implicit Euler,

$$\omega_{i+1} = \omega_i + h_1 f(\omega_{i+1}) \tag{1.87}$$

and subtract (1.86) and (1.87) from (1.10) and (1.38), respectively (assuming $u_i = y_i$), then

$$\begin{aligned} u_{i+1} - y(x_{i+1}) &= -\tfrac{1}{2} h_1^2 f_i' + 0(h_1^3) \\ \omega_{i+1} - y(x_{i+1}) &= \tfrac{1}{2} h_1^2 f_i' + 0(h_1^3) \end{aligned} \tag{1.88}$$

The truncation error can now be estimated by

$$e_{i+1} = \omega_{i+1} - y(x_{i+1}) \simeq \tfrac{1}{2}(\omega_{i+1} - u_{i+1}) \tag{1.89}$$

The process proceeds as follows:

1. Equations (1.86) and (1.87) are used to obtain u_{i+1} and ω_{i+1}.
2. The truncation error is obtained from (1.89).
3. If the truncation error is less than TOL, the step is accepted; if not, the step is repeated.
4. In either case of step(3), the next step-size is calculated according to

$$h_2 = h_1 \left(\frac{\text{TOL}}{e_{i+1}} \right)^{1/2} \tag{1.90}$$

To avoid small errors, one can use an h_2 that is a certain percentage smaller than calculated by (1.90).

Michelsen [16] solved (1.81) with a step-size of h and then again with $h/2$. The semi-implicit algorithm is third-order accurate, so it may be written as

$$\mathbf{u}_{i+1} = \mathbf{y}(x_{i+1}) + \mathbf{g}h^4 + 0(h^5) \tag{1.91}$$

where $\mathbf{g}h^4$ is the dominant, but still unknown, error term. If $\mathbf{u}_{i+1}$ denotes the numerical solution for a step-size of h, and $\boldsymbol{\omega}_{i+1}$ for a step-size of $h/2$, then,

$$\begin{aligned} \mathbf{u}_{i+1} &= \mathbf{y}(x_{i+1}) + \mathbf{g}h^4 + 0(h^5) \\ \boldsymbol{\omega}_{i+1} &= \mathbf{y}(x_{i+1}) + 2\mathbf{g}\left(\frac{h}{2}\right)^4 + 0(h^5) \end{aligned} \tag{1.92}$$

where the $2\mathbf{g}$ in (1.92) accounts for error accumulation in each of the two integration steps. Subtraction of the two equations (1.92) from one another gives

$$\mathbf{e}_{i+1} = \boldsymbol{\omega}_{i+1} - \mathbf{u}_{i+1} = -\tfrac{7}{8}\mathbf{g}h^4 + 0(h^5) \tag{1.93}$$

Provided $\mathbf{e}_{i+1}$ is sufficiently small, the result is accepted. The criterion for step-size acceptance is

$$e^{\{j\}} < \mathrm{TOL}^{\{j\}}, \qquad j = 1, 2, \ldots, m \tag{1.94}$$

where

$$e^{\{j\}} = \text{local truncation error for the } j \text{ component}$$

If this criterion is not satisfied, the step-size is halved and the integration repeated. When integrating stiff problems, this procedure leads to small steps whenever the solution changes rapidly, often times at the start of the integration. As soon as the stiff component has faded away, one observes that the magnitude of $\mathbf{e}$ decreases rapidly and it becomes desirable to increase the step-size. After a successful step with h_i, the step-size h_{i+1} is adjusted by

$$h_{i+1} = h_i \min\left[\left\{4 \max\left|\frac{e^{\{j\}}}{\mathrm{TOL}^{\{j\}}}\right|\right\}^{-1/4}, 3\right], \qquad j = 1, 2, \ldots, m \tag{1.95}$$

For more explanation of (1.95) see [17]. A good discussion of computer algorithms for adjusting the step-size is presented by Johnston [5] and by Krogh [18].

We are now ready to discuss commercial packages that incorporate a variety of techniques for solving systems of IVPs.

MATHEMATICAL SOFTWARE

Most computer installations have preprogrammed computer packages, i.e., software, available in their libraries in the form of subroutines so that they can be accessed by the user's main program. A subroutine for solving IVPs will be designed to compute a numerical solution over $[x_0, x_N]$ and return the value u_N

given x_0, x_N, and u_0. A typical calling sequence could be

```
CALL DRIVE (FUNC, X, XEND, U, TOL),
```

where

FUNC = a user-written subroutine for evaluating $f(x, y)$
X = x_0
XEND = x_N
U = on input contains u_0 and on output contains u_N
TOL = an error tolerance

This is a very simplified call sequence, and more elaborate ones are actually used in commercial routines.

The subroutine DRIVE must contain algorithms that:

1. Implement the numerical integration
2. Adapt the step-size
3. Calculate the local error so as to implement item 2 such that the global error does not surpass TOL
4. Interpolate results to XEND (since h is adaptively modified, it is doubtful that XEND will be reached exactly)

Thus, the creation of a software package, from now on called a code, is a nontrivial problem. Once the code is completed, it must contain sufficient documentation. Several aspects of documentation are significant (from [24]):

1. Comments in the code identifying arguments and providing general instructions to the user (this is valuable because often the code is separated from the other documentation)
2. A document with examples showing how to use the code and illustrating user-oriented aspects of the code
3. Substantial examples of the performance of the code over a wide range of problems
4. Examples showing misuse, subtle and otherwise, of the code and examples of failure of the code in some cases.

Most computer facilities have at least one of the following mathematical libraries:

IMSL [19]
NAG [20]
HARWELL [21]

The Harwell library contains several IVP codes, IMSL has two (which will be discussed below), and NAG contains an extensive collection of routines. These large libraries are not the only sources of codes, and in Table 1.7 we provide a survey of IVP software (excluding IMSL, Harwell, and NAG). Since the production of software has increased tremendously during recent years, any survey of codes will need continual updating. Table 1.7 should provide the reader with an appreciation for the types of codes that are being produced, i.e., the underlying numerical methods. We do not wish to dwell on all of these codes but only to point out a few of the better ones. Recently, a survey of IVP software [33] concluded that RKF45 is the best overall explicit Runge-Kutta routine, while LSODE is quite good for solving stiff problems. LSODE is the update for GEAR/GEARB (versions of which are presently the most used stiff IVP solver) [34].

The comparison of computer codes is a difficult and tricky task, and the results should always be "taken with a grain of salt." Hull et al. [35] have compared nonstiff methods, while Enright et al. [36] compared stiff ones. Although this is an important step, it does not bear directly on how practical a code is. Shampine et al. [37] have shown that how a method is implemented

TABLE 1.7 IVP Codes

Name	Method Implemented	Comments	Reference
RKF45	Runge-Kutta-Fehlberg	—	[22]
GERK	Runge-Kutta-Fehlberg	—	[23]
DE/ODE	Variable-order Adams multistep	DE is limited to 20 equations or less: ODE has no size limit	[6]
DEROOT/ODERT	Variable-order Adams multistep	Same as DE/ODE except that nonlinear scalar equations can be coupled to the IVPs	[6]
GEAR/GEARB	Variable-order Adams multistep and backward multistep	Allow for nonstiff Adams and stiff backward formulas; GEARB allows for banded structure of the Jacobian	[24], [25]
LSODE	—	Replacement for GEAR/GEARB	[26]
EPISODE/EPISODEB	Same as GEAR/GEARB	Differ from GEAR/GEARB in how the variable step-size is performed	[27]
M3RK	Stabilized explicit Runge-Kutta*	Designed to solve systems arising from a method of lines discretization of partial differential equations	[28]
STRIDE	Implicit Runge-Kutta	—	[29]
STIFF3	Semi-implicit Runge-Kutta	See text; Eq. (1.81) with (1.95)	[17]
BLSODE	Blended multistep*	For stiff oscillatory problems	[30]
STINT	Cyclic composite multistep*	—	[31]
SECDER	Variable-order Enright formula*	—	[32]

*Method not covered in this chapter.

may be more important than the choice of method, even when dealing with the best codes. There is a distinction between the best methods and the best codes. In [31] various codes for nonstiff problems are compared, and in [38] GEAR and EPISODE are compared by the authors. One major aspect of code usage that cannot be tested is the user's attitude, including such factors as user time constraints, accessibility of the code, familiarity with the code, etc. It is typically the user's attitude which dictates the code choice for a particular problem, not the question of which is the best code. Therefore, no sophisticated code comparison will be presented. Instead, we illustrate the use of software packages by solving two problems. These problems are chosen to demonstrate the concept of stiffness.

The following codes were used in this study:

1. IMSL-DVERK: Runge-Kutta solver.
2. IMSL-DGEAR: This code is a modified version of GEAR. Two methods are available in this package: a variable-order Adams multistep method and a variable-order implicit multistep method. Implicit methods require Jacobian calculations, and in this package the Jacobian can be (a) user-supplied, (b) internally calculated by finite differences, or (c) internally calculated by a diagonal approximation based on the directional derivative (for more explanation see [24]). The various methods are denoted by the parameter MF, where

MF	Method	Jacobian
10	Adams	—
21	Implicit	User-supplied
22	Implicit	Finite differences
23	Implicit	Diagonal approximation

3. STIFF3: Implements (1.81) using (1.94) and (1.95) to govern the step-size and error.
4. LSODE: updated version of GEAR. The parameter MF is the same as for DGEAR. MF = 23 is not an option in this package.
5. EPISODE: A true variable step-size code based on GEAR. GEAR, DGEAR, and LSODE periodically change the step-size (not on every step) in order to decrease execution time while still maintaining accuracy. EPISODE adapts the step-size on every step (if necessary) and is therefore good for problems that involve oscillations. For decaying or linear problems, EPISODE would probably require larger execution times than GEAR, DGEAR, or LSODE.
6. ODE: Variable-order Adams multistep solver.

We begin our discussions by solving the reactor problem outlined in Example 5:

$$\frac{dy}{dx} = -0.1744 \exp\left[\frac{3.21}{T^*}\right] y$$

$$\frac{dT^*}{dx} = 0.06984 \exp\left[\frac{3.21}{T^*}\right] y \tag{1.96}$$

$$y = T^* = 1 \quad \text{at} \quad x = 0$$

Equations (1.96) are not stiff (see Example 1.5), and all of the codes performed the integration with only minor differences in their solutions. Typical results are shown in Table 1.8. Notice that a decrease in TOL when using DVERK did produce a change in the results (although the change was small). Practically speaking, any of the solutions presented in Table 1.8 would be acceptable. From the discussions presented in this chapter, one should realize that DVERK, ODE, DGEAR (MF = 10), LSODE (MF = 10), and EPISODE (MF = 10) use methods that are capable of solving nonstiff problems, while STIFF3, DGEAR (MF = 21, 22, 23), LSODE (MF = 21, 22), and EPISODE (MF = 21, 22, 23) implement methods for solving stiff systems. Therefore, all of the codes are suitable for solving (1.96). One might expect the stiff problem solvers to require longer execution times because of the Jacobian calculations. This behavior was observed, but since (1.96) is a small system, i.e., two equations, the execution times for all of the codes were on the same order of magnitude. For a larger problem the effect would become significant.

Next, we consider a stiff problem. Robertson [39] originally proposed the

TABLE 1.8 Typical Results from Software Packages Using Eq. (1.96)

	DVERK, TOL = (−4)		DVERK, TOL = (−6)		DGEAR (MF = 21), TOL = (−4)		STIFF3, TOL = (−4)	
x	y	T^*	y	T^*	y	T^*	y	T^*
0.0	1.000000	1.00000	1.000000	1.00000	1.000000	1.00000	1.000000	1.00000
0.1	0.699795	1.12021	0.700367	1.11999	0.700468	1.11994	0.700371	1.11998
0.2	0.528839	1.18868	0.529199	1.18853	0.529298	1.18849	0.529208	1.18853
0.3	0.413483	1.23487	0.413737	1.23477	0.413775	1.23475	0.413745	1.23477
0.4	0.329730	1.26841	0.329919	0.26833	0.329864	1.26836	0.329924	1.26833
0.5	0.266347	1.29379	0.266492	1.29373	0.266349	1.29379	0.266497	1.29373
0.6	0.217094	1.31352	0.217208	1.31347	0.217070	1.31353	0.217211	1.31347
0.7	0.178118	1.32912	0.178209	1.32909	0.178076	1.32914	0.178212	1.32909
0.8	0.146869	1.34164	0.146943	1.34161	0.146801	1.34167	0.146945	1.34161
0.9	0.121569	1.35177	0.121629	1.35175	0.121495	1.35180	0.121630	1.35175
1.0	0.100931	1.36003	0.100980	1.36002	0.100864	1.36006	0.100982	1.36001

following set of differential equations that arise from an autocatalytic reaction pathway:

$$\frac{dy_1}{dt} = -0.04y_1 + 10^4 y_2 y_3$$
$$\frac{dy_2}{dt} = 0.04y_1 - 10^4 y_2 y_3 - 3 \times 10^7 y_2^2 \tag{1.97}$$
$$\frac{dy_3}{dt} = 3 \times 10^7 y_2^2$$
$$y_1(0) = 1, \quad y_2(0) = 0, \quad y_3(0) = 0 \quad \text{at} \quad t = 0$$

The Jacobian matrix is

$$J = \begin{bmatrix} -0.04 & 10^4 y_3 & 10^4 y_2 \\ 0.04 & -10^4 y_3 - 6 \times 10^7 y_2 & -10^4 y_2 \\ 0 & 6 \times 10^7 y_2 & 0 \end{bmatrix} \tag{1.98}$$

When t varies from 0.0 to 0.02, one of the eigenvalues of $\boldsymbol{J}$ changes from -0.04 to $-2{,}450$. Over the complete range of t, $0 \leq t \leq \infty$, one of the eigenvalues varies from -0.04 to -10^4. Figure 1.5 shows the solution of (1.97) for $0 \leq t \leq 10$. Notice the steep gradient in y_2 at small values of t. Thus the problem is very stiff. Caillaud and Padmanabhan [10], Seinfeld et al. [40], Villadsen and Michelsen [17], and Finlayson [41] have discussed this problem. Table 1.9 shows the

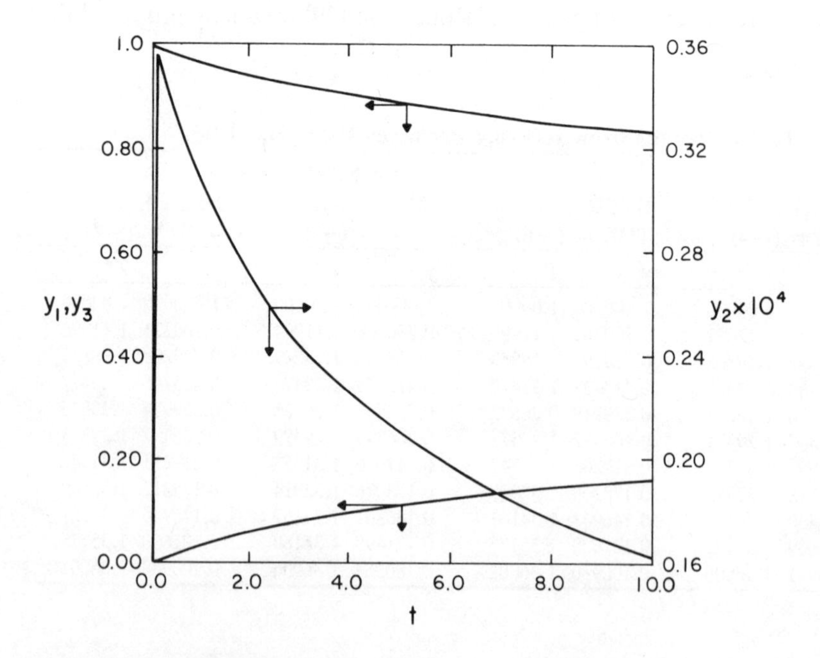

FIGURE 1.5 Results from Eq. (1.97).

results at $t = 10$. At a TOL of 10^{-4} all of the nonstiff methods failed to produce a solution. At smaller tolerance values, the nonstiff methods failed to produce a solution or required excessive execution times, i.e., two orders of magnitude greater than those of the stiff methods. This behavior is due to the fact that the tolerances were too large to achieve a stable solution (recall that the step-size is adapted to meet the error criterion that is governed by the value of TOL) or a solution was obtained but at a high cost (large execution time) because of the very small step-size requirements of nonstiff methods (see section on stiffness).

TABLE 1.9 Comparison of Software Packages on the Robertson Problem (Results at t = 10)

Code	MF	TOL	y_1	$y_2 \times 10^4$	y_3	Execution Time Ratio†
DVERK	—	(−4)	—	No solution	—	—
DVERK	—	(−6)	—	No solution	—	—
DVERK	—	(−8)	—	No solution	—	—
ODE	—	(−4)	—	No solution	—	—
ODE	—	(−6)	0.8411	0.1586	0.1589	339.0
ODE	—	(−9)	0.8414	0.1623	0.1586	347.0
DGEAR	10	(−4)	—	No solution	—	—
DGEAR	21	(−4)	0.8414	0.1624	0.1586	0.25
DGEAR	22	(−4)	0.8414	0.1624	0.1586	1.0
DGEAR	23	(−4)	—	No solution	—	—
DGEAR	10	(−6)	0.8414	0.1619	0.1586	261.0
DGEAR	21	(−6)	0.8414	0.1623	0.1586	1.0
DGEAR	22	(−6)	0.8414	0.1623	0.1586	1.0
DGEAR	23	(−6)	0.8414	0.1624	0.1586	2.5
LSODE	10	(−4)	—	No solution	—	—
LSODE	21	(−4)	—	No solution	—	—
LSODE	22	(−4)	—	No solution	—	—
LSODE‡	10	(−4)	—	No solution	—	—
LSODE‡	21	(−4)	0.8414	0.1623	0.1586	1.75
LSODE‡	22	(−4)	0.8414	0.1623	0.1586	1.75
LSODE	10	(−6)	—	No solution	—	—
LSODE	21	(−6)	0.8414	0.1623	0.1586	1.75
LSODE	22	(−6)	0.8414	0.1623	0.1586	1.75
EPISODE	10	(−4)	—	No solution	—	—
EPISODE	21	(−4)	—	No solution	—	—
EPISODE	22	(−4)	—	No solution	—	—
EPISODE	23	(−4)	—	No solution	—	—
EPISODE	10	(−6)	0.8414	0.1623	0.1586	530.0
EPISODE	21	(−6)	0.8414	0.1623	0.1586	1.5
EPISODE	22	(−6)	0.8414	0.1623	0.1586	1.5
EPISODE	23	(−6)	0.8414	0.1623	0.1586	3.8
STIFF3	—	(−4)	0.8414	0.1623	0.1586	1.25
STIFF3	—	(−6)	0.8414	0.1623	0.1586	3.0
"EXACT"§	—	—	0.841	0.162	0.159	—

†Execution time ratio = execution time/execution time of DGEAR [MF = 21, TOL = (−6)].

‡Tolerance for y_2 is (−8); for y_1 and y_3, (−4).

§Caillaud and Padmanabhan [10].

TABLE 1.10 Comparison of Code Results to the "Exact" Solution for Time = 1, 4, and 10

Code	MF	TOL	t	y_1	$y_2 \times 10^4$	y_3
"EXACT"	—	—	1.0	0.966	0.307	0.0335
			4.0	0.9055	0.224	0.0944
			10.0	0.841	0.162	0.159
STIFF3	—	(−6)	1.0	0.9665	0.3075	0.3351(−1)
			4.0	0.9055	0.2240	0.9446(−1)
			10.0	0.8414	0.1623	0.1586
EPISODE	21	(−6)	1.0	0.9665	0.3075	0.3351(−1)
			4.0	0.9055	0.2240	0.9446(−1)
			10.0	0.8414	0.1623	0.1586
DGEAR	10	(−6)	1.0	0.9665	0.3087	0.3350(−1)
			4.0	0.9055	0.2238	0.9445(−1)
			10.0	0.8414	0.1619	0.1586
DGEAR	21	(−6)	1.0	0.9665	0.3075	0.3351(−1)
			4.0	0.9055	0.2240	0.9446(−1)
			10.0	0.84414	0.1623	0.1586
ODE	—	(−6)	1.0	0.9665	0.3075	0.3351(−1)
			4.0	0.9055	0.2222	0.9452(−1)
			10.0	0.8411	0.1586	0.1589

All of the stiff algorithms were able to produce solutions with execution times on the same order of magnitude. Caillaud and Padmanabhan [10] have studied (1.97) using Runge-Kutta algorithms. Their "exact" results (fourth-order Runge-Kutta with step-size = 0.001) and the results obtained from various codes are presented in Table 1.10. Notice that when a solution was obtained from either a stiff or a nonstiff algorithm, the results were excellent. Therefore, the difference between the stiff and nonstiff algorithms was their execution times.

The previous two examples have illustrated the usefulness of the commercial software packages for the solution of practical problems. It can be concluded that generally one should use a package that incorporates an implicit method for stiff problems and an explicit method for nonstiff problems (this was stated in the section on stiffness, but no examples were given).

We hope to have eliminated the "blackbox" approach to the use of initial-value packages through the illustration of the basic methods and rationale behind the production of these programs. No code is infallible, and when you obtain spurious results from a code, you should be able to rationalize your data with the aid of the code's documentation and the material presented in this chapter.

PROBLEMS*

1. A tubular reactor for a homogeneous reaction has the following dimensions: $L = 2$ m, $R = 0.1$ m. The inlet reactant concentration is $c_0 = 0.03$ kmole/m^3, and the inlet temperature is $T_0 = 700$ K. Other

* See the Preface regarding classes of problems.

data is as follows: $-\Delta H = 10^4$ kJ/kmole, $c_p = 1$ kJ/(kg·K), $E_a = 100$ kJ/kmole, $\rho = 1.2$ kg/m^3, $u_0 = 3$ m/s, and $k_0 = 5s^{-1}$. The appropriate material and energy balance equations are (see [17] for further explanation):

$$\frac{dy}{dz} = -\text{Da}\; y \exp\left[\delta\left(1 - \frac{1}{\theta}\right)\right], \qquad 0 \leq z \leq 1,$$

$$\frac{d\theta}{dz} = \beta\text{Da}\; y \exp\left[\delta\left(1 - \frac{1}{\theta}\right)\right] - H_w(\theta - \theta_w)$$

where

$$\text{Da} = \frac{Lk_0}{u_0}$$

$$\beta = \frac{c_0(-\Delta H)}{\rho c_p T_0}$$

$$\delta = \frac{E_a}{R_g T_0}$$

$$H_w = \frac{2\overline{U}}{R}\left(\frac{L}{\rho c_p u_0}\right)$$

$$y = \frac{c}{c_0}$$

$$\theta = \frac{T}{T_0}$$

If one considers the reactor to be adiabatic, $\overline{U} = 0$, the transport equations can be combined to

$$\frac{d}{dz}(\theta + \beta y) = 0$$

which gives

$$\theta = 1 + \beta(1 - y)$$

using the inlet conditions $\theta = y = 1$. Substitution of this equation into the material balance yields

$$\frac{dy}{dz} = -\text{Da}\; y \exp\left[\frac{\delta\beta(1 - y)}{1 + \beta(1 - y)}\right], \qquad y = 1 \quad \text{at} \quad z = 0$$

(a) Compute y and θ if $\overline{U} = 0$ using an Euler method.

(b) Repeat (a) using a Runge-Kutta method.

(c) Repeat (a) using an implicit method.

(d) Check algorithms (a) to (c) by comparing their solutions to the analytical solution by letting $\delta = 0$.

2. Write a subroutine called EULER such that the call is

```
CALL EULER (FUNC, XO, XOUT, H, TOL, N, Y),
```

where

FUNC = external subroutine to calculate the right-hand-side functions
XO = initial value of the independent variable
XOUT = final value of the independent variable
H = initial step-size
TOL = local error tolerance
N = number of equations to be integrated
Y = vector with N components for the dependent variable y. On input y is the vector initial values, on output it contains the computed values of y at XOUT.

The routine is to perform an Euler integration on

$$\frac{d\mathbf{y}}{dx} = \mathbf{f}(x, \mathbf{y})$$

$$\mathbf{y}(0) = \mathbf{y}_0, \qquad \text{XO} \leq \text{X} \leq \text{XOUT}$$

Create this algorithm such that it contains an error-checking routine and a step-size selection routine. Test your routine by solving Example 5. Hopefully, this problem will give the reader some feel for the difficulty in creating a general-purpose routine.

3.* Repeat Problem 1, but now allow for heat transfer by letting $\overline{U} = 70\ \text{J}/(\text{m}^2\cdot\text{s}\cdot\text{K})$. Locate the position of the hot spot, θ_{max}, with $\theta_w = 1$.

4.* In Example 4 we evaluated a binary batch distillation system. Now consider the same system with recycle (R = recycle ratio) and a constant condenser hold-up M (see Figure 1.6).

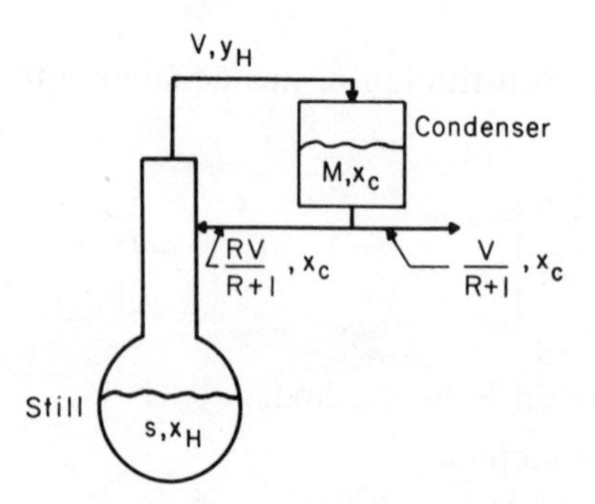

FIGURE 1.6 Batch still with recycle.

A mass balance on *n*-heptane for the condenser is

$$M \frac{dx_c}{dt} = V(y_{\rm H} - x_c)$$

An overall balance on the still gives

$$\frac{ds}{dt} = \frac{-V}{R + 1}$$

while an overall balance on *n*-heptane is

$$s \frac{dx_{\rm H}}{dt} = \frac{-V}{R + 1}(x_c - x_{\rm H})$$

Repeat the calculations of Example 1.4 with $s = 100$, $x_{\rm H} = 0.75$, and $x_c = 0.85$ at $t = 0$. Let $R = 0.3$ and $M = 10$.

5.* Consider the following process where steam passes through a coil, is condensed, and is withdrawn as condensate at the same temperature in order to heat liquid in a well-stirred vessel (see Figure 1.7).
If

F_s = flow rate of steam
H_v = latent heat of vaporization of the steam
F = flow rate of liquid to be heated
T_0 = inlet liquid temperature
T = outlet liquid temperature

and the control valve is assumed to have linear flow characteristics such that instantaneous action occurs, i.e., the only lags in the control scheme

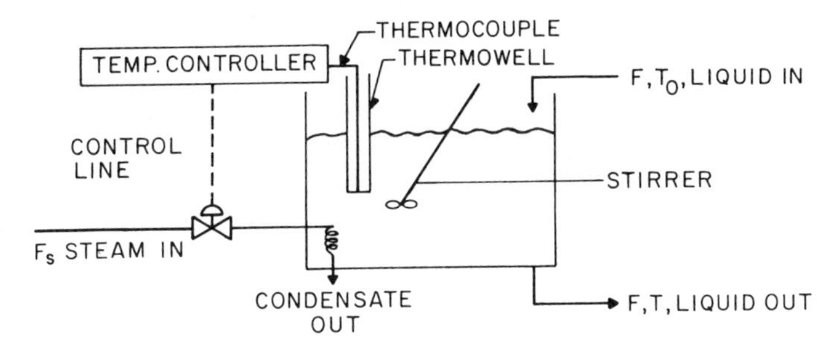

FIGURE 1.7 Temperature control process.

occur in the temperature measures, then the process can be described by

$$Mc_p \frac{dT}{dt} = Fc_p(T_0 - T) + F_s H_v \qquad \text{(liquid energy balance)}$$

$$C_1 \frac{dT_w}{dt} = U_1 A_1 (T - T_w) \qquad \text{(thermowell energy balance)}$$

$$C_2 \frac{dT_t}{dt} = U_2 A_2 (T_w - T_t) \qquad \text{(thermocouple energy balance)}$$

$$F_s = K_p(T_s - T_t) \qquad \text{(proportional control)}$$

For convenience allow

$$\frac{F}{M} = 1 \text{ min}^{-1}$$

$$\frac{H_v}{Mc_p} = 1 \text{ °C/kg}$$

$$T_0 = 50\text{°C}$$

$$10 \frac{U_1 A_1}{C_1} = \frac{U_2 A_2}{C_2} = 1 \text{ min}^{-1}$$

The system of differential equations becomes

$$\frac{dT}{dt} = F_s - T + T_0$$

$$\frac{dT_w}{dt} = 0.1(T - T_w)$$

$$\frac{dT_t}{dt} = T_w - T_t$$

$$F_s = K_p(T_t - T_s)$$

Initially T = 50°C. Investigate the temperature response, $T(t)$, to a 10°C step increase in the designed liquid temperature, T_s = 60°C, for $K_p = 2$ and $K_p = 6$. Recall that with proportional control there is offset in the response.

6.* In a closed system of three components, the following reaction path can occur:

$$A \xrightarrow{k_1} B$$

$$\mathrm{B} + \mathrm{C} \xrightarrow{k_2} \mathrm{A} + \mathrm{C}$$

$$2\mathrm{B} \xrightarrow{k_3} \mathrm{C} + \mathrm{B}$$

with governing differential equations

$$\frac{dC_A}{dt} = -k_1C_A + k_2C_BC_C$$

$$\frac{dC_B}{dt} = k_1C_A - k_2C_BC_C - k_3C_B^2$$

$$\frac{dC_C}{dt} = k_3C_B^2$$

$$C_A(0) = 1, \qquad C_B(0) = C_C(0) = 0$$

Calculate the reaction pathway for $k_1 = 0.08$, $k_2 = 2 \times 10^4$, and $k_3 = 6 \times 10^7$.

7. Develop a numerical procedure to solve

$$\frac{d^2f}{dr^2} + \frac{2}{r}\frac{df}{dr} = \Phi^2 R(f), \qquad 0 \leqslant r \leqslant 1$$

$$\frac{df}{dr}(0) = 0, \qquad f(1) = 1$$

Hint: Let $df/dr(1) = \alpha$ and choose α to satisfy $(df/dr)(0) = 0$.

(Later in this text we will discuss this method for solving boundary-value problems. Methods of this type are called shooting methods.)

REFERENCES

1. Conte, S. D., and C. deBoor, *Elementary Numerical Analysis: An Algorithmic Approach,* 3rd Ed., McGraw-Hill, New York (1980).
2. Kehoe, J. P. G., and J. B. Butt, "Interactions of Inter- and Intraphase Gradients in a Diffusion Limited Catalytic Reaction," A.I.Ch.E. J., *18,* 347 (1972).
3. Price, T. H., and J. B. Butt, "Catalyst Poisoning and Fixed Bed Reactor Dynamics-II," Chem. Eng. Sci., *32,* 393 (1977).
4. Gear, C. W., *Numerical Initial-Value Problems in Ordinary Differential Equations,* Prentice-Hall, Englewood Cliffs, N.J. (1971).
5. Johnston, R. L., *Numerical Methods—A Software Approach,* Wiley, New York (1982).
6. Shampine, L. F., and M. K. Gordon, *Computer Solution of Ordinary Differential Equations: The Initial Value Problem,* Freeman, San Francisco (1975).

7. Rosenbrock, H. H., "Some General Implicit Processes for the Numerical Solution of Differential Equations," Comput. J., *5,* 329 (1963).
8. Calahan, D., "Numerical Considerations in the Transient Analysis and Optimal Design of Nonlinear Circuits," Digest Record of Joint Conference on Mathematical and Computer Aids to Design, ACM/SIAM/IEEE, Anaheim, Calif. 129 (1969).
9. Allen, R. H., "Numerically Stable Explicit Integration Techniques Using a Linearized Runge-Kutta Extension," Boeing Scientific Laboratories Document D1-82-0929 (1969).
10. Caillaud, J. B., and L. Padmanabhan, "An Improved Semi-implicit Runge-Kutta Method for Stiff Systems," Chem. Eng. J., *2,* 227 (1971).
11. Norsett, S. P., "One-Step Methods of Hermite-type for Numerical Integration of Stiff Systems," BIT, *14,* 63 (1974).
12. Alexander, R., "Diagonally Implicit Runge-Kutta Methods for Stiff ODES," SIAM J. Numer. Anal., *14,* 1006 (1977).
13. Bui, T. D., and T. R. Bui, "Numerical Methods for Extremely Stiff Systems of Ordinary Differential Equations," Appl. Math. Modelling, *3,* 355 (1979).
14. Burka, M. K., "Solution of Stiff Ordinary Differential Equations by Decomposition and Orthogonal Collocation," A.I.Ch.E.J., *28,* 11 (1982).
15. Lambert, J. D., "Stiffness," in *Computational Techniques for Ordinary Differential Equations,* I. Gladwell, and D. K. Sayers (eds.), Academic, London (1980).
16. Michelsen, M. L., "An Efficient General Purpose Method for the Integration of Stiff Ordinary Differential Equations," A.I.Ch.E.J., *22,* 594 (1976).
17. Villadsen, J., and M. L. Michelsen, *Solution of Differential Equation Models by Polynomial Approximation,* Prentice-Hall, Englewood Cliffs, N.J. (1978).
18. Krogh, F. T., "Algorithms for Changing Step Size," SIAM J. Numer. Anal., *10,* 949 (1973).
19. International Mathematics and Statistics Libraries Inc., Sixth Floor–NBC Building, 7500 Bellaire Boulevard, Houston, Tex.
20. Numerical Algorithms Group (USA) Inc., 1250 Grace Court, Downers Grove, Ill.
21. Harwell Subroutine Libraries, Computer Science and Systems Division of the United Kingdom Atomic Energy Authority, Harwell, England.
22. Forsythe, G. E., M. A. Malcolm, and C. B. Moler, *Computer Methods for Mathematical Computations,* Prentice-Hall, Englewood Cliffs, N.J. (1977).
23. Shampine, L. F., and H. A. Watts, "Global Error Estimation for Ordinary Differential Equations," ACM TOMS, *2,* 172 (1976).

24. Hindmarsh, A. C., "GEAR: Ordinary Differential Equation System Solver," Lawrence Livermore Laboratory Report UCID-30001 (1974).

25. Hindmarsh, A. C., "GEARB: Solution of Ordinary Differential Equations Having Banded Jacobians," Lawrence Livermore Laboratory Report UCID-30059 (1975).

26. Hindmarsh, A. C., "LSODE and LSODI, Two New Initial Value Ordinary Differential Equation Solvers," ACM SIGNUM Newsletter December (1980).

27. Byrne, G. D., and A. C. Hindmarsh, "EPISODEB: An Experimental Package for the Integration of Systems of Ordinary Differential Equations with Banded Jacobians," Lawrence Livermore Laboratory Report UCID-30132 (1976).

28. Verwer, J. G., "Algorithm 553. M3RK, An Explicit Time Integrator for Semidiscrete Parabolic Equations," ACM TOMS, *6,* 236 (1980).

29. Butcher, J. C., K. Burrage, and F. H. Chipman, "STRIDE Stable Runge-Kutta Integrator for Differential Equations," Report Series No. 150, Department of Mathematics, University of Auckland, New Zealand (1979).

30. Skeel, R., and A. Kong, "Blended Linear Multistep Methods," ACM TOMS, *3,* 326 (1977).

31. Tendler, J. M., T. A. Bickart, and Z. Picel, "Algorithm 534. STINT: STiff INTegrator," ACM TOMS, *4,* 399 (1978).

32. Addison, C. A., "Implementing a Stiff Method Based Upon the Second Derivative Formulas," University of Toronto Department of Computer Science, Technical Report No. 130/79 (1979).

33. Gladwell, I., J. A. I. Craigie, and C. R. Crowther, "Testing Initial-Value Problem Subroutines as Black Boxes," Numerical Analysis Report No. 34, Department of Mathematics, University of Manchester, Manchester, England.

34. Gaffney, P. W., "Information and Advice on Numerical Software," Oak Ridge National Laboratory Report No. ORNL/CSD/TM-147, May (1981).

35. Hull, T. E., W. H. Enright, B. M. Fellen, and A. E. Sedgwick, "Comparing Numerical Methods for Ordinary Differential Equations," SIAM J. Numer. Anal., *9,* 603 (1972).

36. Enright, W. H., T. E. Hull, and B. Lindberg, "Comparing Numerical Methods for Stiff Systems of ODEs," BIT, *15,* 10 (1975).

37. Shampine, L. F., H. A. Watts, and S. M. Davenport, "Solving Nonstiff Ordinary Differential Equations—The State of the Art," SIAM Rev., *18,* 376 (1976).

38. Byrne, G. D., A. C. Hindmarsh, K. R. Jackson, and H. G. Brown, "A Comparison of Two ODE CODES: GEAR and ERISODE," Comput. Chem. Eng., *1,* 133 (1977).

39. Robertson, A. H., "Solution of a Set of Reaction Rate Equations," in *Numerical Analysis,* J. Walsh (ed.), Thomson Brook Co., Washington (1967).

40. Seinfeld, J. H., L. Lapidus, and M. Hwang, "Review of Numerical Integration Techniques for Stiff Ordinary Differential Equations," Ind. Eng. Chem. Fund., *9,* 266 (1970).

41. Finlayson, B. A., *Nonlinear Analysis in Chemical Engineering,* McGraw-Hill, New York (1980).

BIBLIOGRAPHY

Only a brief overview of IVPs has been given in this chapter. For additional or more detailed information, see the following:

Finlayson, B. A., *Nonlinear Analysis in Chemical Engineering,* McGraw-Hill, New York (1980).

Forsythe, G. E., M. A. Malcolm, and C. B. Moler, *Computer Methods for Mathematical Computations,* Prentice-Hall, Englewood Cliffs, N.J. (1977).

Gear, C. W., *Numerical Initial-Value Problems in Ordinary Differential Equations,* Prentice-Hall, Englewood Cliffs, N.J. (1971).

Hall, G., and J. M. Watt (eds.), *Modern Numerical Methods for Ordinary Differential Equations,* Clarendon Press, Oxford (1976).

Johnston, R. L., *Numerical Methods—A Software Approach,* Wiley, New York (1982).

Lambert, J. D., *Computational Methods in Ordinary Differential Equations,* Wiley, New York (1973).

Shampine, L. F., and M. K. Gordon, *Computer Solution of Ordinary Differential Equations: The Initial Value Problem,* Freeman, San Francisco (1975).

Villadsen, J., and M. L. Michelsen, *Solution of Differential Equation Models by Polynomial Approximation,* Prentice-Hall, Englewood Cliffs, N.J. (1978).

2

Boundary-Value Problems for Ordinary Differential Equations: Discrete Variable Methods

INTRODUCTION

In this chapter we discuss discrete variable methods for solving BVPs for ordinary differential equations. These methods produce solutions that are defined on a set of discrete points. Methods of this type are initial-value techniques, i.e., shooting and superposition, and finite difference schemes. We will discuss initial-value and finite difference methods for linear and nonlinear BVPs, and then conclude with a review of the available mathematical software (based upon the methods of this chapter).

BACKGROUND

One of the most important subdivisions of BVPs is between linear and nonlinear problems. In this chapter linear problems are assumed to have the form

$$y' = F(\mathrm{x})\mathbf{y} + \mathbf{z}(\mathrm{x}), \qquad a < x < b \tag{2.1a}$$

with

$$A\ \mathbf{y}(a) + B\ \mathbf{y}(b) = \boldsymbol{\gamma} \tag{2.1b}$$

where $\boldsymbol{\gamma}$ is a constant vector, and nonlinear problems to have the form

$$\mathbf{y}' = \mathbf{f}(x,\mathbf{y}), \qquad a < x < b \tag{2.2a}$$

with

$$\mathbf{g}(\mathbf{y}(a), \mathbf{y}(b)) = \mathbf{0} \tag{2.2b}$$

If the number of differential equations in systems (2.1a) or (2.2a) is n, then the number of independent conditions in (2.1b) and (2.2b) is n.

In practice, few problems occur naturally as first-order systems. Most are posed as higher-order equations that can be converted to a first-order system. All of the software discussed in this chapter require the problem to be posed in this form.

Equations (2.1b) and (2.2b) are called boundary conditions (BCs) since information is provided at the ends of the interval, i.e., at $x = a$ and $x = b$. The conditions (2.1b) and (2.2b) are called nonseparated BCs since they can involve a combination of information at $x = a$ and $x = b$. A simpler situation that frequently occurs in practice is that the BCs are separated; that is, (2.1b) and (2.2b) can be replaced by

$$\boldsymbol{A}\,\mathbf{y}(a) = \boldsymbol{\gamma}_1, \qquad \boldsymbol{B}\,\mathbf{y}(b) = \boldsymbol{\gamma}_2 \tag{2.3}$$

where $\boldsymbol{\gamma}_1$ and $\boldsymbol{\gamma}_2$ are constant vectors, and

$$\mathbf{g}_1(\mathbf{y}\,(a)) = \mathbf{0}, \qquad \mathbf{g}_2(\mathbf{y}\,(b)) = \mathbf{0} \tag{2.4}$$

respectively, where the total number of independent conditions remains equal to n.

INITIAL-VALUE METHODS

Shooting Methods

We first consider the single linear second-order equation

$$Ly \equiv -y'' + p(x)y' + q(x)y = r(x), \qquad a < x < b \tag{2.5a}$$

with the general linear two-point boundary conditions

$$\begin{aligned} a_0 y(a) - a_1 y'(a) &= \alpha \\ b_0 y(b) + b_1 y'(b) &= \beta \end{aligned} \tag{2.5b}$$

where a_0, a_1, α, b_0, b_1, and β are constants, such that

$$\begin{aligned} &a_0 a_1 \geqslant 0, \qquad |a_0| + |a_1| \neq 0 \\ &b_0 b_1 \geqslant 0, \qquad |b_0| + |b_1| \neq 0 \\ &|a_0| + |b_0| \neq 0 \end{aligned} \tag{2.5c}$$

We assume that the functions $p(x)$, $q(x)$, and $r(x)$ are continuous on $[a, b]$ and that $q(x) > 0$. With this assumption [and (2.5c)] the solution of (2.5) is unique

[1]. To solve (2.5) we first define two functions, $y^{(1)}(x)$ and $y^{(2)}(x)$, on $[a, b]$ as solutions of the respective initial-value problems

$$Ly^{(1)} = r(x), \qquad y^{(1)}(a) = -\alpha C_1, \qquad y^{(1)\prime}(a) = -\alpha C_0 \tag{2.6a}$$

$$Ly^{(2)} = 0, \qquad y^{(2)}(a) = a_1, \qquad y^{(2)\prime}(a) = a_0 \tag{2.6b}$$

where C_0 and C_1 are any constants such that

$$a_1 C_0 - a_0 C_1 = 1 \tag{2.7}$$

The function $y(x)$ defined by

$$y(x) \equiv y(x; s) = y^{(1)}(x) + s y^{(2)}(x), \qquad a \leqslant x \leqslant b \tag{2.8}$$

satisfies $a_0 y(a) - a_1 y'(a) = \alpha(a_1 C_0 - a_0 C_1) = \alpha$, and will be a solution of (2.5) if s is chosen such that

$$\phi(s) = b_0 y(b; s) + b_1 y'(b; s) - \beta = 0 \tag{2.9}$$

This equation is linear in s and has the single root

$$s = \frac{\beta - [b_0 y^{(1)}(b) + b_1 y^{(1)\prime}(b)]}{[b_0 y^{(2)}(b) + b_1 y^{(2)\prime}(b)]} \tag{2.10}$$

Therefore, the method involves:

1. Converting the BVP into an IVP by specifying extra initial conditions	**1.** (2.5) to (2.6)
2. Guessing the initial conditions and solving the IVP over the entire interval	**2.** Guess C_0, evaluate C_1 from (2.7), and solve (2.6)
3. Solving for s and constructing y.	**3.** Evaluate (2.10) for s; use s in (2.8)

The shooting method consists in simply carrying out the above procedure numerically; that is, compute approximations to $y^{(1)}(x)$, $y^{(1)\prime}(x)$, $y^{(2)}(x)$, $y^{(2)\prime}(x)$ and use them in (2.8) and (2.10). To solve the initial-value problems (2.6), first write them as equivalent first-order systems:

$$\begin{bmatrix} w^{(1)} \\ v^{(1)} \end{bmatrix}' = \begin{bmatrix} v^{(1)} \\ pv^{(1)} + qw^{(1)} - r \end{bmatrix}$$
$$w^{(1)}(a) = -\alpha C_1, \qquad v^{(1)}(a) = -\alpha C_0 \tag{2.11}$$

and

$$\begin{bmatrix} w^{(2)} \\ v^{(2)} \end{bmatrix}' = \begin{bmatrix} v^{(2)} \\ pv^{(2)} + qw^{(2)} \end{bmatrix}$$
$$w^{(2)}(a) = a_1, \qquad v^{(2)}(a) = a_0 \tag{2.12}$$

respectively. Now any of the methods discussed in Chapter 1 can be employed to solve (2.11) and (2.12).

Let the numerical solutions of (2.11) and (2.12) be

$$W_i^{(1)},\ V_i^{(1)},\ W_i^{(2)},\ V_i^{(2)}, \qquad i = 0, 1, \ldots, N \tag{2.13}$$

respectively, for

$$x_i = a + ih, \qquad i = 0, 1, \ldots, N$$

$$h = \frac{b - a}{N}$$

At the point $x_0 = a$, the exact data can be used so that

$$\begin{aligned} W_0^{(1)} &= -\alpha C_1, & W_0^{(2)} &= -\alpha C_0 \\ V_0^{(1)} &= a_1, & V_0^{(2)} &= a_0 \end{aligned} \tag{2.14}$$

To approximate the solution $y(x)$, set

$$Y_i = W_i^{(1)} + SW_i^{(2)} \tag{2.15}$$

where

$$Y_i \simeq y(x_i)$$

$$S = \frac{\beta - [b_0 W_N^{(1)} + b_1 V_N^{(1)}]}{[b_0 W_N^{(2)} + b_1 V_N^{(2)}]} \tag{2.16}$$

This procedure can work well but is susceptible to round-off errors. If $W_i^{(1)}$ and $W_i^{(2)}$ in (2.15) are nearly equal and of opposite sign for some range of i values, cancellation of the leading digits in Y_i can occur.

Keller [1] posed the following example to show how cancellation of digits can occur. Suppose that the solution of the IVP (2.6) grows in magnitude as $x \to b$ and that the boundary condition at $x = b$ has $b_1 = 0$ [$y(b) = \beta$ is specified]. Then if $|\beta| << |b_0 W_N^{(1)}|$

$$S \simeq -\frac{W_N^{(1)}}{W_N^{(2)}} \tag{2.17}$$

and

$$Y_i \simeq W_i^{(1)} - \left[\frac{W_N^{(1)}}{W_N^{(2)}}\right] W_i^{(2)} \tag{2.18}$$

Clearly the cancellation problem occurs here for x_i near b. Note that the solution $W_i^{(1)}$ need not grow very fast, and in fact for $\beta = 0$ the difficulty is always potentially present. If the loss of significant digits cannot be overcome by the use of double precision arithmetic, then multiple-shooting techniques (discussed later) can be employed.

We now consider a second-order nonlinear equation of the form

$$y'' = f(x, y, y'), \qquad a < x < b \tag{2.19a}$$

subject to the general two-point boundary conditions

$$\begin{aligned} a_0 y(a) - a_1 y'(a) &= \alpha, \qquad a_i \geq 0 \\ b_0 y(b) + b_1 y'(b) &= \beta, \qquad b_i \geq 0 \\ a_0 + b_0 &> 0 \end{aligned} \tag{2.19b}$$

The related IVP is

$$u'' = f(x, u, u'), \qquad a < x < b \tag{2.20a}$$

$$\begin{aligned} u(a) &= a_1 s - c_1 \alpha \\ u'(a) &= a_0 s - c_0 \alpha \end{aligned} \tag{2.20b}$$

where

$$a_1 c_0 - a_0 c_1 = 1$$

The solution of (2.20), $u = u(x; s)$, will be a solution of (2.19) if s is a root of

$$\phi(s) = b_0 u(b; s) + b_1 u'(b; s) - \beta = 0 \tag{2.21}$$

To carry out this procedure numerically, convert (2.20) into a first-order system:

$$\begin{bmatrix} w \\ v \end{bmatrix}' = \begin{bmatrix} v \\ f(x, w, v) \end{bmatrix} \tag{2.22a}$$

with

$$\begin{aligned} w(a) &= a_1 s - c_1 \alpha \\ v(a) &= a_0 s - c_0 \alpha \end{aligned} \tag{2.22b}$$

In order to find s, one can apply Newton's method to (2.21), giving

$$\begin{aligned} s^{[k+1]} &= s^{[k]} - \frac{\phi(s^{[k]})}{\phi'(s^{[k]})}, \qquad k = 0, 1, \ldots \\ s^{[0]} &= \text{arbitrary} \end{aligned} \tag{2.23}$$

To find $\phi'(s)$, first define

$$\xi(x) = \frac{\partial w(x; s)}{\partial s} \quad \text{and} \quad \eta(x) = \frac{\partial v(x; s)}{\partial s} \tag{2.24}$$

Differentiation of (2.22) with respect to s gives

$$\begin{aligned} \xi' &= \eta, & \xi(a) &= a_1 \\ \eta' &= \frac{\partial f}{\partial v}\eta + \frac{\partial f}{\partial w}\xi, & \eta(a) &= a_0 \end{aligned} \tag{2.25}$$

Solution of (2.25) allows for calculation of ϕ' as

$$\phi' = b_0\,\xi(b;s) + b_1\,\eta(b;s) \tag{2.26}$$

Therefore, the numerical solution of (2.25) would be computed along with the numerical solution of (2.22). Thus, one iteration in Newton's method (2.23) requires the solution of two initial-value problems.

EXAMPLE 1

An important problem in chemical engineering is to predict the diffusion and reaction in a porous catalyst pellet. The goal is to predict the overall reaction rate of the catalyst pellet. The conservation of mass in a spherical domain gives

$$D\left[\frac{1}{r^2}\frac{d}{dr}\left(r^2\frac{dc}{dr}\right)\right] = k\mathscr{R}(c), \qquad 0 < r < r_p$$

where

$$\begin{aligned} r &= \text{radial coordinate } (r_p = \text{pellet radius}) \\ D &= \text{diffusivity} \\ c &= \text{concentration of a given chemical} \\ k &= \text{rate constant} \\ \mathscr{R}(c) &= \text{reaction rate function} \end{aligned}$$

with

$$\frac{dc}{dr} = 0 \quad \text{at} \quad r = 0 \qquad \text{(symmetry about the origin)}$$

$$c = c_0 \quad \text{at} \quad r = r_p \qquad \text{(concentration fixed at surface)}$$

If the pellet is isothermal, an energy balance is not necessary. We define the effectiveness factor E as the average reaction rate in the pellet divided by the average reaction rate if the rate of reaction is evaluated at the surface. Thus

$$E = \frac{\displaystyle\int_0^{r_p} \mathscr{R}(c(r))r^2\,dr}{\displaystyle\int_0^{r_p} \mathscr{R}(c_0)r^2\,dr}$$

We can integrate the mass conservation equation to obtain

$$D\int_0^{r_p}\left[\frac{1}{r^2}\frac{d}{dr}\left(r^2\frac{dc}{dr}\right)\right]r^2\,dr = k\int_0^{r_p}\mathscr{R}(c)r^2\,dr = Dr_p^2\left.\frac{dc}{dr}\right|_{r_p}$$

Hence the effectiveness factor can be rewritten as

$$E = \frac{3r_p^2\,D\left.\dfrac{dc}{dr}\right|_{r_p}}{k\mathscr{R}(c_0)}$$

If $E = 1$, then the overall reaction rate in the pellet is equal to the surface value and mass transfer has no limiting effects on the reaction rate. When $E < 1$, then mass transfer effects have limited the overall rate in the pellet; i.e., the average reaction rate in the pellet is lower than the surface value of the reaction rate because of the effects of diffusion.

Now consider a sphere (5 mm in diameter) of γ-alumina upon which Pt is dispersed in order to catalyze the dehydrogenation of cyclohexane. At 700 K, the rate constant k is 4 s^{-1}, and the diffusivity D is 5×10^{-2} cm^2/s. Set up the equations necessary to calculate the concentration profile of cyclohexane within the pellet and also the effectiveness factor for a general $\mathscr{R}(c)$. Next, solve these equations for $\mathscr{R}(c) = c$, and compare the results with the analytical solution.

SOLUTION

Define

$$C = \frac{\text{concentration of cyclohexane}}{\text{concentration of cyclohexane at the surface of the sphere}}$$

R = dimensionless radial coordinate based on the radius of the sphere ($r_p = 2.5$ mm)

Assume that the spherical pellet is isothermal. The conservation of mass equation for cyclohexane is

$$\frac{d^2C}{dR^2} + \frac{2}{R}\frac{dC}{dR} = \Phi^2 \frac{\mathscr{R}(c)}{c_0}, \qquad 0 < R < 1,$$

with

$$\frac{dC}{dR} = 0 \quad \text{at} \quad R = 0 \qquad \text{(due to symmetry)}$$

$$C = 1 \quad \text{at} \quad R = 1 \qquad \text{(by definition)}$$

where

$$\Phi = r_p\sqrt{\frac{k}{D}} \qquad \text{(Thiele modulus)}$$

Since $\mathscr{R}(c)$ is a general function of c, it may be nonlinear in c. Therefore, assume that $\mathscr{R}(c)$ is nonlinear and rewrite the conservation equation in the form of (2.19):

$$\frac{d^2C}{dR^2} = \Phi^2 \frac{\mathscr{R}(c)}{c_0} - \frac{2}{R}\frac{dC}{dR} = f(R, C, C')$$

The related IVP systems become

$$\begin{bmatrix} w \\ v \end{bmatrix}' = \begin{bmatrix} v \\ \Phi^2 \dfrac{\mathscr{R}(c)}{c_0} - \dfrac{2}{R} v \end{bmatrix}, \quad \begin{bmatrix} \xi \\ \eta \end{bmatrix}' = \begin{bmatrix} \eta \\ \Phi^2 \dfrac{d}{dw}\left(\dfrac{\mathscr{R}(c)}{c_0}\right) - \dfrac{2}{R}\eta \end{bmatrix}$$

with

$$w(0) = s, \qquad \xi(0) = 1$$

$$v(0) = 0, \qquad \eta(0) = 0$$

and

$$\phi(s) = w(1; s) - 1$$

$$\phi'(s) = \xi(1; s)$$

Choose $s^{[0]}$, and solve the above system of equations to obtain a solution. Compute a new s by

$$s^{[k+1]} = s^{[k]} - \frac{w(1; s^{[k]}) - 1}{\xi(1; s^{[k]})}, \qquad k = 0, 1, \ldots$$

and repeat until convergence is achieved.

Using the data provided, we get $\Phi = 2.236$. If $\mathscr{R}(c) = c$, then the problem is linear and no Newton iteration is required. The IMSL routine DVERK (see Chapter 1) was used to integrate the first-order system of equations. The results, along with the analytical solution calculated from [2],

$$C = \frac{\sinh(\Phi R)}{R \sinh(\Phi)}$$

are shown in Table 2.1. Notice that the computed results are the same as the analytical solution (to four significant figures). In Table 2.1 we also compare

TABLE 2.1 Results from Example 1
TOL = 10^{-6} for DVERK

R	C, Analytical Solution	C, Computed Solution ($s = 0.4835$)
0.0	0.4835	0.4835
0.2	0.4998	0.4998
0.4	0.5506	0.5506
0.6	0.6422	0.6422
0.8	0.7859	0.7859
1.0	1.0000	1.0000
E	0.7726	0.7727

the computed value of E, which is defined as

$$E = \frac{3 \left.\dfrac{dC}{dR}\right|_1}{\Phi^2}$$

with the analytical value from [2],

$$E = \frac{3}{\Phi}\left[\frac{1}{\tanh(\Phi)} - \frac{1}{\Phi}\right]$$

Again, the results are quite good.

Physically, one would expect the concentration of cyclohexane to decrease as R decreases since it is being consumed by reaction. Also, notice that the concentration remains finite at $R = 0$. Therefore, the reaction has not gone to completion in the center of the catalytic pellet. Since $E < 1$, the average reaction rate in the pellet is less than the surface value, thus showing the effects of mass transfer.

EXAMPLE 2

If the system described in Example 1 remains the same except for the fact that the reaction rate function now is second-order, i.e., $\mathcal{R}(c) = c^2$, compute the concentration profile of cyclohexane and calculate the value of the effectiveness factor. Let $c_0 = 1$.

SOLUTION

The material balance equation is now

$$\frac{d^2C}{dR^2} + \frac{2}{R}\frac{dC}{dR} = \Phi^2 C^2, \qquad 0 < R < 1$$

$$\frac{dC}{dR} = 0 \quad \text{at} \quad R = 0$$

$$C = 1 \quad \text{at} \quad R = 1$$

$$\Phi = 2.236$$

The related IVP systems are

$$\begin{bmatrix} w \\ v \end{bmatrix}' = \begin{bmatrix} v \\ \Phi^2 w^2 - \dfrac{2}{R} v \end{bmatrix}, \qquad \begin{bmatrix} \zeta \\ \eta \end{bmatrix}' = \begin{bmatrix} \eta \\ 2\Phi^2 w\zeta - \dfrac{2}{R} \eta \end{bmatrix}$$

with

$$w(0) = s, \qquad \xi(0) = 1$$

$$v(0) = 0, \qquad \eta(0) = 0$$

and

$$\phi(s) = w(1; s) - 1$$

$$\phi'(s) = \xi(1; s)$$

The results are shown in Table 2.2. Notice the effect of the tolerances set for DVERK (TOLD) and on the Newton iteration (TOLN). At TOLN $= 10^{-3}$, the convergence criterion was not sufficiently small enough to match the boundary condition at $R = 1.0$. At TOLN $= 10^{-6}$ the boundary condition at $R = 1$ was achieved. Decreasing either TOLN or TOLD below 10^{-6} produced the same results as shown for TOLN = TOLD $= 10^{-6}$.

In the previous two examples, the IVPs were not stiff. If a stiff IVP arises in a shooting algorithm, then a stiff IVP solver, for example, LSODE (MF = 21), would have to be used to perform the integration.

Systems of BVPs can be solved by initial-value techniques by first converting them into an equivalent system of first-order equations. Consider the system

$$\mathbf{y}' = \mathbf{f}(x, \mathbf{y}), \qquad a < x < b \tag{2.27a}$$

with

$$A\,\mathbf{y}(a) + B\,\mathbf{y}(b) = \boldsymbol{\alpha} \tag{2.27b}$$

or more generally

$$\mathbf{g}(\mathbf{y}(a), \mathbf{y}(b)) = \mathbf{0} \tag{2.27c}$$

The associated IVP is

$$\mathbf{u}' = \mathbf{f}(x, \mathbf{u}) \tag{2.28a}$$

$$\mathbf{u}(a) = \mathbf{s} \tag{2.28b}$$

where

$$\mathbf{s} = \text{vector of unknowns}$$

TABLE 2.2 Results from Example 2

R	C, TOLD $= 10^{-3}$† TOLN $= 10^{-3}$‡	C, TOLD $= 10^{-6}$ TOLN $= 10^{-3}$	C, TOLD $= 10^{-6}$ TOLN $= 10^{-6}$
0.0	0.5924	0.5924	0.5921
0.2	0.6042	0.6042	0.6039
0.4	0.6415	0.6415	0.6411
0.6	0.7101	0.7101	0.7096
0.8	0.8220	0.8220	0.8214
1.0	1.0008	1.0008	1.0000
E	0.6752	0.6752	0.6742
s	0.5924	0.5924	0.5921

† Tolerance for DVERK.

‡ Tolerance on Newton iteration.

We now seek $\mathbf{s}$ such that $\mathbf{u}(x; \mathbf{s})$ is a solution of (2.27). This occurs if $\mathbf{s}$ is a root of the system

$$\boldsymbol{\phi}(\mathbf{s}) = A\,\mathbf{s} + B\,\mathbf{u}(b; \mathbf{s}) - \boldsymbol{\alpha} = \mathbf{0} \tag{2.29}$$

or more generally

$$\boldsymbol{\phi}(\mathbf{s}) = \mathbf{g}(\mathbf{s}, \mathbf{u}(b; \mathbf{s})) = \mathbf{0} \tag{2.30}$$

Thus far we have only discussed shooting methods that "shoot" from $x = a$. Shooting can be applied in either direction. If the solutions of the IVP grow from $x = a$ to $x = b$, then it is likely that the shooting method will be most effective in reverse, that is, using $x = b$ as the initial point. This procedure is called reverse shooting.

Multiple Shooting

Previously we have discussed some difficulties that can arise when using a shooting method. Perhaps the best known difficulty is the loss in accuracy caused by the growth of solutions of the initial-value problem. Multiple shooting attempts to prevent this problem. Here, we outline multiple-shooting methods that are used in software libraries.

Multiple shooting is designed to reduce the growth of the solutions of the IVPs that must be solved. This is done by partitioning the interval into a number of subintervals, and then simultaneously adjusting the "initial" data in order to satisfy the boundary conditions and appropriate continuity conditions. Consider a system of n first-order equations of the form (2.27), and partition the interval as

$$a = x_0 < x_1 < \ldots < x_{N-1} < x_N = b \tag{2.31}$$

Define

$$\begin{aligned} h_i &= x_i - x_{i-1} \\ t &= \frac{x - x_{i-1}}{h_i} \\ \boldsymbol{\tau}_i(t) &= \mathbf{y}(x) = \mathbf{y}(x_{i-1} + th_i) \\ \mathbf{r}_i(t, \boldsymbol{\tau}_i) &= h_i\,\mathbf{f}(x_{i-1} + th_i, \boldsymbol{\tau}_i) \end{aligned} \tag{2.32}$$

for

$$i = 1, 2, \ldots, N$$

With this change of variables, (2.27) becomes

$$\frac{d\boldsymbol{\tau}_i}{dt} = \mathbf{r}_i(t, \mathbf{r}_i), \qquad 0 < t < 1 \tag{2.33}$$

for

$$i = 1, 2, \ldots, N$$

The boundary conditions are now

$$A\,\boldsymbol{\tau}_1(0) + B\,\boldsymbol{\tau}_N(1) = \boldsymbol{\alpha} \qquad \text{[for (2.27b)]} \tag{2.34a}$$

or

$$\mathbf{g}(\boldsymbol{\tau}_1(0), \boldsymbol{\tau}_N(1)) = \mathbf{0} \qquad \text{[for (2.27c)]} \tag{2.34b}$$

In order to have a continuous solution to (2.27), we require

$$\boldsymbol{\tau}_{i+1}(0) - \boldsymbol{\tau}_i(1) = \mathbf{0}, \qquad i = 1, 2, \ldots, N-1 \tag{2.35}$$

The N systems of n first-order equations can thus be written as

$$\frac{d}{dt}\boldsymbol{\psi} = \mathcal{R}(t, \boldsymbol{\psi}) \tag{2.36}$$

with

$$P\,\boldsymbol{\psi}(0) + Q\,\boldsymbol{\psi}(1) = \boldsymbol{\gamma}$$

or

$$\mathbf{G} = \mathbf{0}$$

where

$$\begin{aligned}
\boldsymbol{\psi} &= [\boldsymbol{\tau}_1(t), \boldsymbol{\tau}_2(t), \ldots, \boldsymbol{\tau}_N(t)]^T \\
\mathcal{R}(t, \boldsymbol{\psi}) &= [\mathbf{r}_1(t, \boldsymbol{\tau}_1), \mathbf{r}_2(t, \boldsymbol{\tau}_2), \ldots, \mathbf{r}_N(t, \boldsymbol{\tau}_N)]^T \\
\boldsymbol{\gamma} &= [\boldsymbol{\alpha}, \mathbf{0}, \ldots, \mathbf{0}]^T \\
\mathbf{0} &= [\mathbf{0}, \mathbf{0}, \ldots, \mathbf{0}]^T
\end{aligned}$$

$$P = \begin{bmatrix} A & & & & \\ & I & & & \\ & & \cdot & \mathbf{0} & \\ & \mathbf{0} & & \cdot & \\ & & & & \cdot \\ & & & & I \end{bmatrix}$$

$$Q = \begin{bmatrix} \mathbf{0} & & & B \\ -I & \mathbf{0} & & \\ & \cdot & \cdot & \\ & & \cdot & \cdot \\ & & & \cdot \quad \cdot \\ & & -I & \mathbf{0} \end{bmatrix}$$

$$\mathbf{G} = \begin{bmatrix} \boldsymbol{\tau}_2(0) - \boldsymbol{\tau}_1(1) \\ \boldsymbol{\tau}_3(0) - \boldsymbol{\tau}_2(1) \\ \vdots \\ \boldsymbol{\tau}_N(0) - \boldsymbol{\tau}_{N-1}(1) \\ \mathbf{g}(\boldsymbol{\tau}_1(0), \boldsymbol{\tau}_N(1)) \end{bmatrix}$$

The related IVP problem is

$$\frac{d}{dt}\mathbf{U} = \mathcal{R}(t, \mathbf{U}), \qquad 0 < t < 1 \tag{2.37}$$

with

$$\mathbf{U}(0) = \mathbf{S}$$

where

$$\mathbf{S} = [\mathbf{S}_1, \mathbf{S}_2, \ldots, \mathbf{S}_N]^T$$

$$\mathbf{U} = [\mathbf{u}_1, \mathbf{u}_2, \ldots, \mathbf{u}_N]^T$$

The solution of (2.37) is a solution to (2.36) if **S** is a root of

$$\mathbf{\Phi}(\mathbf{S}) = \boldsymbol{P}\,\mathbf{S} + \boldsymbol{Q}\,\mathbf{U}(1; \mathbf{S}) - \boldsymbol{\gamma} = \mathbf{0}$$

or (2.38)

$$\mathbf{\Phi}(\mathbf{S}) = \mathbf{G} = \mathbf{0}$$

depending on whether the BCs are of form (2.27b) or (2.27c). The solution procedure consists of first guessing the "initial" data **S**, then applying ordinary shooting on (2.37) while also performing a Newton iteration on (2.38). Obviously, two major considerations are the mesh selection, i.e., choosing x_i, $i = 1, \ldots, N - 1$, and the starting guess for **S**. These difficulties will be discussed in the section on software.

An alternative shooting procedure would be to integrate in both directions up to certain matching points. Formally speaking, this method includes the previous method as a special case. It is not clear a priori which method is preferable [3].

Superposition

Another initial-value method is called superposition and is discussed in detail by Scott and Watts [4]. We will outline the method for the following linear equation

$$\mathbf{y}'(x) = \boldsymbol{F}(x)\mathbf{y}(x) + \mathbf{g}(x), \qquad a < x < b \tag{2.39a}$$

with

$$\boldsymbol{A}\,\mathbf{y}(a) = \boldsymbol{\alpha} \tag{2.39b}$$

$$\boldsymbol{B}\,\mathbf{y}(b) = \boldsymbol{\beta}$$

The technique consists of finding a solution $\mathbf{y}(x)$ such that

$$\mathbf{y}(x) = \mathbf{v}(x) + \boldsymbol{U}(x)\mathbf{c} \tag{2.40}$$

where the matrix U satisfies

$$U'(x) = F(x)U(x) \tag{2.41a}$$

$$A\,U(a) = \mathbf{0} \tag{2.41b}$$

the vector $\mathbf{v}(x)$ satisfies

$$\mathbf{v}'(x) = F(x)\,\mathbf{v}(x) + \mathbf{g}(x) \tag{2.42a}$$

$$\mathbf{v}(a) = \boldsymbol{\alpha} \tag{2.42b}$$

and the vector of constants $\mathbf{c}$ is chosen to satisfy the boundary conditions at $x = b$:

$$B\,U(b)\mathbf{c} = -B\,\mathbf{v}(b) + \boldsymbol{\beta} \tag{2.43}$$

The matrix $U(x)$ is often referred to as the fundamental solution, and the vector $\mathbf{v}(x)$ the particular solution.

In order for the method to yield accurate results, $\mathbf{v}(x)$ and the columns of $U(x)$ must be linearly independent [5]. The initial conditions (2.41b) and (2.42b) theoretically ensure independence; however, due to the finite world length used by computers, the solutions may lose their numerical independence (see [5] for full explanation). When this happens, the resulting matrix problem (2.43) may give inaccurate answers for $\mathbf{c}$. Frequently, it is impossible to extend the precision of the computations in order to overcome this difficulty. Therefore, the basic superposition method must be modified.

Analogous to using multiple shooting to overcome the difficulties with shooting methods, one can modify the superposition method by subdividing the interval as in (2.31), and then defining a superposition solution on each subinterval by

$$\begin{aligned} \mathbf{y}_i(x) &= \mathbf{v}_i(x) + U_i(x)\mathbf{c}_i(x), \qquad x_{i-1} \leqslant x \leqslant x_i \\ i &= 1, 2, \ldots, N, \end{aligned} \tag{2.44}$$

where

$$U_i'(x) = F(x)\,U_i(x) \tag{2.45}$$

$$U_i(x_{i-1}) = U_{i-1}(x_{i-1}), \qquad A\,U_1(a) = \mathbf{0}$$

$$\mathbf{v}_i'(x) = F(x)\mathbf{v}_i(x) + \mathbf{g}(x) \tag{2.46}$$

$$\mathbf{v}_i(x_{i-1}) = \mathbf{v}_{i-1}(x_{i-1}), \qquad \mathbf{v}_1(a) = \boldsymbol{\alpha}$$

and

$$\mathbf{y}_i(x_i) = \mathbf{y}_{i+1}(x_i) \tag{2.47}$$

$$B\,U_N(b)\mathbf{c}_N = -B\,\mathbf{v}_N(b) + \boldsymbol{\beta} \tag{2.48}$$

The principle of the method is then to piece together the solutions defined on the various subintervals to obtain the desired solution. At each of the mesh points x_i the linear independence of the solutions must be checked. One way to guarantee independence of solutions over the entire interval is to keep them nearly orthogonal. Therefore, the superposition algorithm must be coupled with a routine that checks for orthogonality of the solutions, and each time the vectors start to lose their linear independence, they must be orthonormalized [4,5] to regain linear independence. Obviously, one of the major problems in implementing this method is the location of the orthonormalization points x_i.

Nonlinear problems can also be solved using superposition, but they first must be "linearized." Consider the following nonlinear BVP:

$$\begin{aligned} \mathbf{y}'(x) &= \mathbf{f}(x, \mathbf{y}), \qquad a < x < b \\ A\,\mathbf{y}(a) &= \boldsymbol{\alpha} \\ B\,\mathbf{y}(b) &= \boldsymbol{\beta} \end{aligned} \tag{2.49}$$

If Newton's method is applied directly to the nonlinear function $\mathbf{f}(x, \mathbf{y})$, then the method is called quasilinearization. Quasilinearization of (2.49) gives

$$\mathbf{y}'_{(k+1)}(x) = \mathbf{f}(x, \mathbf{y}_{(k)}(x)) + J(x, \mathbf{y}_{(k)}(x))(\mathbf{y}_{(k+1)}(x) - \mathbf{y}_{(k)}(x)), \quad k = 0, 1, \ldots \tag{2.50}$$

where

$$\begin{aligned} J(x, \mathbf{y}_{(k)}(x)) &= \text{Jacobian of } \mathbf{f}(x, \mathbf{y}_{(k)}(x)) \\ k &= \text{iteration number} \end{aligned}$$

One iteration of (2.50) can be solved by the superposition methods outlined above since it is a linear system.

FINITE DIFFERENCE METHODS

Up to this point, we have discussed initial-value methods for solving boundary-value problems. In this section we cover finite difference methods. These methods are said to be global methods since they simultaneously produce a solution over the entire interval.

The basic steps for a finite difference method are as follows: first, choose a mesh on the interval of interest, that is, for $[a,b]$

$$a = x_0 < x_1 < \ldots < x_N < x_{N+1} = b \tag{2.51}$$

such that the approximate solution will be sought at these mesh points; second, form the algebraic equations required to satisfy the differential equation and the BCs by replacing derivatives with difference quotients involving only the mesh points; and last, solve the algebraic system of equations.

Linear Second-Order Equations

We first consider the single linear second-order equation

$$Ly \equiv -y'' + p(x)y' + q(x)y = r(x), \qquad a < x < b \tag{2.52a}$$

subject to the Dirichlet boundary conditions

$$\begin{aligned} y(a) &= \alpha \\ y(b) &= \beta \end{aligned} \tag{2.52b}$$

On the interval $[a, b]$ impose a uniform mesh,

$$x_i = a + ih, \qquad i = 0, 1, \ldots, N + 1,$$

$$h = \frac{b - a}{N + 1}$$

The parameter h is called the mesh-size, and the points x_i are the mesh points. If $y(x)$ has continuous derivatives of order four, then, by Taylor's theorem,

$$y(x+h) = y(x) + hy'(x) + \frac{h^2}{2!}y''(x) + \frac{h^3}{3!}y'''(x) + \frac{h^4}{4!}y''''(\xi),$$

$$x_i \leq \xi \leq x_i + h \tag{2.53}$$

$$y(x-h) = y(x) - hy'(x) + \frac{h^2}{2!}y''(x) - \frac{h^3}{3!}y'''(x) + \frac{h^4}{4!}y''''(\bar{\xi}),$$

$$x_i - h \leq \bar{\xi} \leq x_i \tag{2.54}$$

From (2.53) and (2.54) we obtain

$$y'(x) = \left[\frac{y(x + h) - y(x)}{h}\right] - \frac{h}{2!}y''(x) - \frac{h^2}{3!}y'''(x) - \frac{h^3}{4!}y''''(\xi) \tag{2.55}$$

$$y'(x) = \left[\frac{y(x) - y(x - h)}{h}\right] + \frac{h}{2!}y''(x) - \frac{h^2}{3!}y'''(x) + \frac{h^3}{4!}y''''(\bar{\xi}) \tag{2.56}$$

respectively. The forward and backward difference equations (2.55) and (2.56) can be written as

$$y'(x_i) = \frac{y_{i+1} - y_i}{h} + 0(h) \tag{2.57}$$

$$y'(x_i) = \frac{y_i - y_{i-1}}{h} + 0(h) \tag{2.58}$$

respectively, where

$$y_i = y(x_i)$$

Thus, Eqs. (2.57) and (2.58) are first-order accurate difference approximations to the first derivative. A difference approximation for the second derivative is

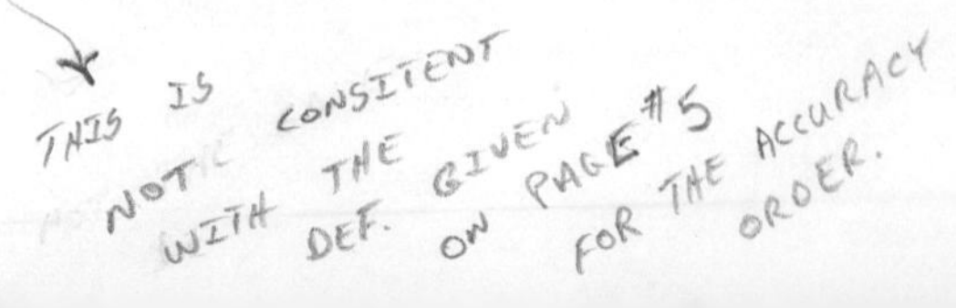

obtained by adding (2.54) and (2.53) to give

$$y(x + h) + y(x - h) = 2y(x) + h^2y''(x) + \frac{h^4}{4!}[y''''(\xi) + y''''(\bar{\xi})] \tag{2.59}$$

from which we obtain

$$y''(x_i) = \frac{(y_{i+1} - 2y_i + y_{i-1})}{h^2} + 0(h^2) \tag{2.60}$$

If the BVP under consideration contains both first and second derivatives, then one would like to have an approximation to the first derivative compatible with the accuracy of (2.60). If (2.54) is subtracted from (2.53), then

$$2hy'(x) = y(x + h) - y(x - h) - \frac{h^3}{3!}y'''(x) + \frac{h^4}{4!}[y''''(\bar{\xi}) - y''''(\xi)] \tag{2.61}$$

and hence

$$y'(x_i) = \left[\frac{y_{i+1} - y_{i-1}}{2h}\right] + 0(h^2) \tag{2.62}$$

which is the central difference approximation for the first derivative and is clearly second-order accurate.

To solve the given BVP, at each *interior* mesh point x_i we replace the derivatives in (2.52a) by the corresponding second-order accurate difference approximations to obtain

$$L_hu_i \equiv -\left[\frac{u_{i+1} - 2u_i + u_{i-1}}{h^2}\right] + p(x_i)\left[\frac{u_{i+1} - u_{i-1}}{2h}\right] + q(x_i)u_i = r(x_i)$$

$$i = 1, \ldots, N \tag{2.63}$$

and

$$u_0 = \alpha, \qquad u_{N+1} = \beta$$

where

$$u_i \simeq y_i$$

The result of multiplying (2.63) by $h^2/2$ is

$$\frac{h^2}{2}L_hu_i = a_iu_{i-1} + b_iu_i + c_iu_{i+1} = \frac{h^2}{2}r(x_i), \qquad i = 1, 2, \ldots, N$$

$$u_0 = \alpha, \qquad u_{N+1} = \beta \tag{2.64}$$

where

$$a_i = -\frac{1}{2}\left[1 + \frac{h}{2}p(x_i)\right]$$

$$b_i = \left[1 + \frac{h^2}{2}q(x_i)\right]$$

$$c_i = -\frac{1}{2}\left[1 - \frac{h}{2}p(x_i)\right]$$

The system (2.64) in vector notation is

$$A\,\mathbf{u} = \mathbf{r} \tag{2.65}$$

where

$$\mathbf{u} = [u_1, u_2, \ldots, u_N]^T$$

$$\mathbf{r} = \frac{h^2}{2}\left[r(x_1) - \frac{2a_1\alpha}{h^2}, r(x_2), \ldots, r(x_{N-1}) - \frac{2c_N\beta}{h^2}\right]^T$$

$$A = \begin{bmatrix} b_1 & c_1 & & & & 0 \\ a_2 & b_2 & c_2 & & & \\ & \cdot & \cdot & \cdot & & \\ & & \cdot & \cdot & \cdot & \\ & & & \cdot & \cdot & \cdot \\ & & & a_{N-1} & b_{N-1} & c_{N-1} \\ 0 & & & & a_N & b_N \end{bmatrix}$$

A matrix of the form A is called a tridiagonal. This special form permits a very efficient application of the Gaussian elimination procedure (described in Appendix C).

To estimate the error in the numerical solution of BVPs by finite difference methods, first define the local truncation errors $\tau_i[\emptyset]$ in L_h as an approximation of L, for any smooth function $\emptyset(x)$ by

$$\tau_i[\emptyset] = L_h\emptyset(x_i) - L\emptyset(x_i), \qquad i = 1, 2, \ldots, N \tag{2.66}$$

If $\emptyset(x)$ has continuous fourth derivatives in $[a, b]$, then for L defined in (2.52) and L_h defined in (2.63),

$$\tau_i[\emptyset] = -\left[\frac{\emptyset(x_i + h) - 2\emptyset(x_i) + \emptyset(x_i - h)}{h^2}\right] + \emptyset''(x_i)$$
$$+ p(x_i)\left[\frac{\emptyset(x_i + h) - \emptyset(x_i - h)}{2h} - \emptyset'(x_i)\right] \tag{2.67}$$

or by using (2.59) and (2.61),

$$\tau_i[\emptyset] = -\frac{h^2}{4!}[\emptyset''''(\gamma_i) - 2p(x_i)\emptyset'''(\bar{\gamma}_i)], \qquad i = 1, \ldots, N \tag{2.68}$$

$$x_{i-1} \leq \gamma_i \leq x_{i+1}, \qquad x_{i-1} \leq \bar{\gamma}_i \leq x_{i+1}$$

From (2.67) we find that L_h is consistent with L, that is, $\tau_i[\emptyset] \to 0$ as $h \to 0$, for all functions $\emptyset(x)$ having continuous second derivatives on $[a, b]$. Further, from (2.68), it is apparent that L_h has second-order accuracy (in approximating L) for functions $\emptyset(x)$ with continuous fourth derivatives on $[a, b]$. For sufficiently small h, L_h is *stable*, i.e., for all functions v_i, $i = 0, 1, \ldots, N + 1$ defined on x_i, $i = 0, 1, \ldots, N + 1$, there is a positive constant M such that

$$|v_i| \leq M\{\max(|v_0|, |v_{N+1}|) + \max_{1 \leq i \leq N} |L_h v_i|\}$$

for $i = 0, 1, \ldots, N + 1$. If L_h is stable and consistent, it can be shown that the error is given by

$$|u_i - y(x_i)| \leqslant M \max_{1 \leqslant j \leqslant i} |\tau_j[y]|, \qquad i = 1, \ldots, N \tag{2.69}$$

(for proof see Chapter 3 of [1]).

Flux Boundary Conditions

Consider a one-dimensional heat conduction problem that can be described by Fourier's law and is written as

$$\frac{1}{z^s}\frac{d}{dz}\left[z^s k \frac{dT}{dz}\right] = g(z), \qquad 0 < z < 1 \tag{2.70}$$

where

k = thermal conductivity

$g(z)$ = heat generation or removal function

s = geometric factor: 0, rectangular; 1, cylindrical; 2, spherical

In practical problems, boundary conditions involving the flux of a given component occur quite frequently. To illustrate the finite difference method with flux boundary conditions, consider (2.70) with $s = 0$, $g(z) = z$, k = constant, and

$$T = T_0 \quad \text{at} \quad z = 0 \tag{2.71}$$

$$\frac{dT}{dz} + \lambda_1 T = \lambda_2 \quad \text{at} \quad z = 1 \tag{2.72}$$

where λ_1 and λ_2 are given constants. Since the governing differential equation is (2.70), the difference formula is

$$u_{i+1} - 2u_i + u_{i-1} = \frac{h^2}{k} z_i, \qquad i = 1, 2, \ldots, N \tag{2.73a}$$

with

$$u_0 = T_0 \tag{2.73b}$$

$$\frac{dT}{dz} + \lambda_1 T = \lambda_2 \quad \text{at} \quad z = 1 \tag{2.73c}$$

Since u_{N+1} is now an unknown, a difference equation for (2.73c) must be determined in order to solve for u_{N+1}.

To determine u_{N+1}, first introduce a "fictitious" point x_{N+2} and a corresponding value u_{N+2}. A second-order correct approximation for the first derivative at $z = 1$ is

$$\frac{dT}{dz} \simeq \frac{T_{N+2} - T_N}{2h} \tag{2.74}$$

Therefore, approximate (2.73c) by

$$\frac{u_{N+2} - u_N}{2h} + \lambda_1 u_{N+1} = \lambda_2 \tag{2.75}$$

and solve for u_{N+2}

$$u_{N+2} = 2h(\lambda_2 - \lambda_1 u_{N+1}) + u_N \tag{2.76}$$

The substitution of (2.76) into (2.73a) with $i = N + 1$ gives

$$(\lambda_2 - \lambda_1 u_{N+1})h - u_{N+1} + u_N = \frac{h^2}{2k} \tag{2.77}$$

Notice that (2.77) contains two unknowns, u_N and u_{N+1}, and together with the other $i = 1, 2, \ldots, N$ equations of type (2.73a), maintains the tridiagonal structure of the matrix $\mathbf{A}$. This method of dealing with the flux condition is called the method of fictitious boundaries for obvious reasons.

EXAMPLE 3

A simple but practical application of heat conduction is the calculation of the efficiency of a cooling fin. Such fins are used to increase the area available for heat transfer between metal walls and poorly conducting fluids such as gases. A rectangular fin is shown in Figure 2.1. To calculate the fin efficiency one must first calculate the temperature profile in the fin. If $L >> B$, no heat is lost from the end or from the edges, and the heat flux at the surface is given by

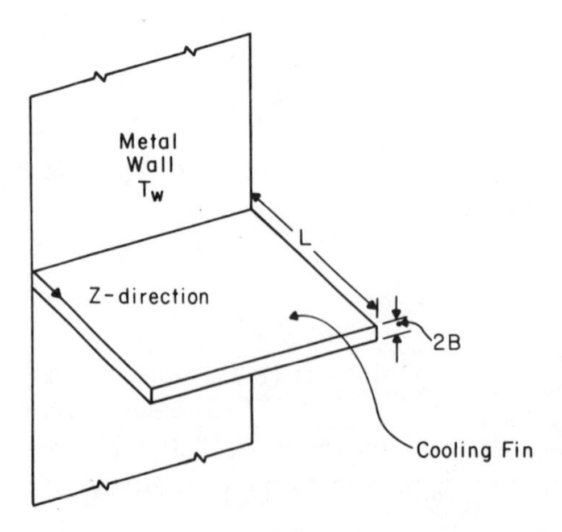

FIGURE 2.1 Cooling fin.

$q = \eta(T - Ta)$ in which the convective heat transfer coefficient η is constant as is the surrounding fluid temperature Ta, then the governing differential equation is

$$\frac{d^2T}{dz^2} = \frac{\eta}{kB}(T - Ta)$$

where

$$k = \text{thermal conductivity of the fin}$$

and

$$T(0) = T_w$$

$$\frac{dT}{dz}(L) = 0$$

Calculate the temperature profile in the fin, and demonstrate the order of accuracy of the finite difference method.

SOLUTION

Define

$$\theta = \frac{T - Ta}{T_w - Ta}$$

$$x = \frac{z}{L}$$

$$H = \sqrt{\frac{\eta L^2}{kB}}$$

The problem can be reformulated as

$$\frac{d^2\theta}{dx^2} = H^2\theta, \qquad \theta(0) = 1, \qquad \frac{d\theta}{dx}(1) = 0$$

The analytical solution to the governing differential equation is

$$\theta = \frac{\cosh H(1 - x)}{\cosh H}$$

For this problem the finite difference method (2.63) becomes

$$a_i u_{i-1} + b_i u_i + c_i u_{i+1} = 0, \qquad i = 1, 2, \ldots, N$$

where

$$a_i = 1$$
$$c_i = 1$$
$$b_i = -(2 + h^2H^2)$$

with

$$u_0 = 1$$

and

$$2u_N - (2 + h^2H^2)u_{N+1} = 0$$

Numerical results are shown in Table 2.3. Physically, one would expect θ to decrease as x increases since heat is being removed from the fin by convection. From these results we demonstrate the order of accuracy of the finite difference method.

If the error in approximation is $0(h^P)$ [see (2.68)], then an estimate of P can be determined as follows. If $u_j(h)$ is the approximate solution calculated using a mesh-size h and

$$e_j(h) = y(x_j) - u_j(h), \qquad j = 1, \ldots, N + 1$$

with

$$\|e(h)\| = \max_j |y(x_j) - u_j(h)|$$

then let

$$\|e(h)\| \simeq \psi h^P$$

where ψ is a constant. Use a sequence of h values, that is, $h_1 > h_2 > \ldots$, and write

$$\|e(h_t)\| = \psi h_t^P$$

TABLE 2.3 Results of $(d^2\theta)/(dx^2) = 4\theta$, $\theta(0) = 1$, $\theta'(1) = 0$

x	Analytical solution	θ, $h = 0.2$	Error†, $h = 0.2$	Error, $h = 0.1$	Error, $h = 0.05$	Error, $h = 0.02$
0.0	1.00000	1.00000	—	—	—	—
0.2	0.68509	0.68713	2.0 (−3)	5.1 (−4)	1.2 (−4)	2.0 (−5)
0.4	0.48127	0.48421	2.9 (−3)	7.4 (−4)	1.8 (−4)	2.9 (−5)
0.6	0.35549	0.35876	3.2 (−3)	8.2 (−4)	2.0 (−4)	3.3 (−5)
0.8	0.28735	0.29071	3.3 (−3)	8.4 (−4)	2.1 (−4)	3.4 (−5)
1.0	0.26580	0.26917	3.3 (−3)	8.5 (−4)	2.1 (−4)	3.4 (−5)

† Error = θ − analytical solution.

The value of P can be determined as

$$P = \frac{\ln\left[\dfrac{\|e(h_{t-1})\|}{\|e(h_t)\|}\right]}{\ln\left[\dfrac{h_{t-1}}{h_t}\right]}$$

Using the data in Table 2.3 gives:

t	h_t	$\|e(h_t)\|$	$\ln\left[\dfrac{\|e_{t-1}\|}{\|e_t\|}\right]$	$\ln\left[\dfrac{h_{t-1}}{h_t}\right]$	P
1	0.20	3.3 (−3)	—	—	—
2	0.10	8.5 (−4)	1.356	0.693	1.96
3	0.05	2.1 (−4)	1.398	0.693	2.01
4	0.02	3.4 (−5)	1.820	0.916	1.99

One can see the second-order accuracy from these results.

Integration Method

Another technique can be used for deriving the difference equations. This technique uses integration, and a complete description of it is given in Chapter 6 of [6].

Consider the following differential equation

$$-\frac{d}{dx}\left[\omega(x)\frac{dy}{dx}\right] + p(x)\frac{dy}{dx} + q(x)y = r(x)$$

$$a < x < b \tag{2.78}$$

$$\alpha_1 y(a) - \beta_1 y'(a) = \gamma_1$$

$$\alpha_2 y(b) + \beta_2 y'(b) = \gamma_2$$

where $\omega(x), p(x), q(x)$, and $r(x)$ are only piecewise continuous and hence possess a number of jump discontinuities. Physically, such problems arise from steady-state diffusion problems for heterogeneous materials, and the points of discontinuity represent interfaces between successive homogeneous compositions. For such problems y and $\omega(x)y'$ must be continuous at an interface $x = \eta$, that is,

$$y(\eta^-) = y(\eta^+)$$

$$\omega(\eta^+)\left.\frac{dy}{dx}\right|_{\eta^+} = \omega(\eta^-)\left.\frac{dy}{dx}\right|_{\eta^-} \tag{2.79}$$

$$a < \eta < b$$

Choose any set of points $a = x_0 < x_1 < \ldots < x_{N+1} = b$ such that the discontinuities of ω, p, q, and r are a subset of these points, that is, $\eta = x_i$ for some i. Note that the mesh spacings $h_i = x_{i+1} - x_i$ need not be uniform. Integrate (2.78) over the interval $x_i \leqslant x \leqslant x_i + h_i/2 \equiv x_{i+1/2}$, $1 \leqslant i \leqslant N$, to give:

$$-\omega_{i+1/2}\frac{dy_{i+1/2}}{dx} + \omega(x_i^+)\frac{dy(x_i^+)}{dx} + \int_{x_i}^{x_{i+1/2}} p(x)\frac{dy}{dx}\,dx + \int_{x_i}^{x_{i+1/2}} y(x)q(x)\,dx = \int_{x_i}^{x_{i+1/2}} r(x)\,dx \tag{2.80}$$

We can also integrate (2.78) over the interval $x_{i-1/2} \leqslant x \leqslant x_i$ to obtain:

$$-\omega(x_i^-)\frac{dy(x_i^-)}{dx} + \omega_{i-1/2}\frac{dy_{i-1/2}}{dx} + \int_{x_{i-1/2}}^{x_i} p(x)\frac{dy}{dx}\,dx + \int_{x_{i-1/2}}^{x_i} y(x)q(x)\,dx = \int_{x_{x-1/2}}^{x_i} r(x)\,dx \tag{2.81}$$

Adding (2.81) and (2.80) and employing (2.79) gives

$$-\omega_{i+1/2}\frac{dy_{i+1/2}}{dx} + \omega_{i-1/2}\frac{dy_{i-1/2}}{dx} + \int_{x_{i-1/2}}^{x_{i+1/2}} p(x)\frac{dy}{dx}\,dx + \int_{x_{i-1/2}}^{x_{i+1/2}} y(x)q(x)\,dx = \int_{x_{i-1/2}}^{x_{i+1/2}} r(x)\,dx \tag{2.82}$$

The derivatives in (2.82) can be approximated by central differences, and the integrals can be approximated by

$$\int_{x_{i-1/2}}^{x_{i+1/2}} g(x)\,dx = \int_{x_{i-1/2}}^{x_i} g(x)\,dx + \int_{x_i}^{x_{i+1/2}} g(x)\,dx \simeq g_i^-\left(\frac{h_{i-1}}{2}\right) + g_i^+\left(\frac{h_i}{2}\right) \tag{2.83}$$

where

$$g_i^- = g(x_i^-)$$
$$g_i^+ = g(x_i^+)$$

Using these approximations in (2.82) results in

$$-\omega_{i+1/2}\left[\frac{u_{i+1} - u_i}{h_i}\right] + \omega_{i-1/2}\left[\frac{u_i - u_{i-1}}{h_{i-1}}\right] + p_i^+\left[\frac{u_{i+1} - u_i}{2}\right] + p_i^-\left[\frac{u_i - u_{i-1}}{2}\right] + u_i\left[\frac{q_i^- h_{i-1} + q_i^+ h_i}{2}\right] = \left[\frac{r_i^- h_{i-1} + r_i^+ h_i}{2}\right],$$
$$1 \leqslant i \leqslant N \tag{2.84}$$

At the left boundary condition, if $\beta_1 = 0$, then $u_0 = \gamma_1/\alpha_1$. If $\beta_1 > 0$, then u_0 is unknown. In this case, direct substitution of the boundary condition into (2.80) for $i = 0$ gives

$$-\omega_{1/2}\left[\frac{u_1 - u_0}{h_0}\right] + \omega_0\left[\frac{\alpha_1 u_0 - \gamma_1}{\beta_1}\right] + p_0\left[\frac{u_1 - u_0}{2}\right] + q_0 u_0 \frac{h_0}{2} = \frac{r_0 h_0}{2} \tag{2.85}$$

The treatment of the right-hand boundary is straightforward. Thus, these expressions can be written in the form

$$-L_i u_{i-1} + D_i u_i - U_i u_{i+1} = R_i, \qquad i = 1, 2, \ldots, N$$

where

$$L_i = \frac{\omega_{i-1/2}}{h_{i-1}} + \frac{p_i^-}{2} \tag{2.86}$$

$$U_i = \frac{\omega_{i+1/2}}{h_i} - \frac{p_i^+}{2}$$

$$D_i = L_i + U_i + \frac{q_i^- h_{i-1} + q_i^+ h_i}{2}$$

$$R_i = \frac{r_i^- h_{i-1} + r_i^+ h_i}{2}$$

Again, if $\beta_1 > 0$, then

$$L_0 = 0$$

$$U_0 = \frac{\omega_{1/2}}{h_0} - \frac{p_0}{2} \tag{2.87}$$

$$D_0 = L_0 + U_0 + \frac{\omega_0 \alpha_1}{\beta_1} + \frac{q_0 h_0}{2}$$

$$R_0 = \frac{r_0 h_0}{2} + \frac{\omega_0 \gamma_1}{\beta}$$

Summarizing, we have a system of equations

$$\mathbf{A\,u} = \mathbf{R} \tag{2.88}$$

where $\mathbf{A}$ is an $m \times m$ tridiagonal matrix where $m = N, N + 1$, or $N + 2$ depending upon the boundary conditions; for example $m = N + 1$ for the combination of one Dirichlet condition and one flux condition.

EXAMPLE 4

A nuclear fuel element consists of a cylindrical core of fissionable material surrounded by a metal cladding. Within the fissionable material heat is produced as a by-product of the fission reaction. A single fuel element is pictured in Figure 2.2. Set up the difference equations in order to calculate the radial temperature profile in the element.

Data: Let

$$T_c = T(r_c)$$
$$k_f = \text{thermal conductivity of core, } k_f \neq k_f(r)$$
$$k_c = \text{thermal conductivity of the cladding, } k_c \neq k_c(r)$$
$$S(r) = \text{source function of thermal energy, } S = 0 \text{ for } r > r_f$$

SOLUTION

Finite Difference Formulation

The governing differential equation is:

$$\frac{d}{dr}\left(rk\frac{dT}{dr}\right) = rS$$

$$\frac{dT}{dr} = 0 \quad \text{at} \quad r = 0$$

$$T = T_c \quad \text{at} \quad r = r_c$$

with

$$S = \begin{cases} S(r), & 0 \leq r \leq r_f \\ 0, & r > r_f \end{cases}$$

and

$$k_f\frac{dT}{dr}\bigg|_{r_f^-} = k_c\frac{dT}{dr}\bigg|_{r_f^+}$$

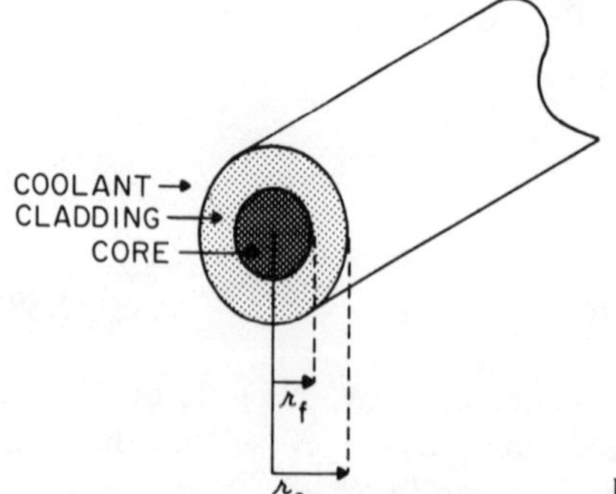

FIGURE 2.2 Nuclear fuel element.

By using (2.84), the difference formula becomes

$$-r_{i+\frac{1}{2}}k\left[\frac{u_{i+1}-u_i}{h_i}\right]+r_{i-\frac{1}{2}}k\left[\frac{u_i-u_{i-1}}{h_{i-1}}\right]=\left[\frac{h_{i-1}+h_i}{2}\right]r_iS_i$$

If $i = 0$ is the center point and $i = j$ is the point r_f, then the system of difference equations becomes

$$u_1 - u_0 = 0$$

$$-\left(\frac{r_{i+\frac{1}{2}}k_f}{h_i}\right)u_{i+1}+k_f\left[\frac{r_{i+\frac{1}{2}}}{h_i}+\frac{r_{i-\frac{1}{2}}}{h_{i-1}}\right]u_i-\left(\frac{r_{i-\frac{1}{2}}k_f}{h_{i-1}}\right)u_{i-1}=\frac{S_i}{2}[h_{i-1}r_i+h_i r_i],$$

$$i=1,\ldots,j-1$$

$$-\left(\frac{r_{j+\frac{1}{2}}k_c}{h_j}\right)u_{j+1}+\left[\frac{r_{j+\frac{1}{2}}k_c}{h_j}+\frac{r_{j-\frac{1}{2}}k_f}{h_{j-1}}\right]u_j-\left(\frac{r_{j-\frac{1}{2}}k_f}{h_{j-1}}\right)u_{j-1}=\frac{1}{2}[h_{j-1}r_jS_j],\ i=j$$

$$-\left(\frac{r_{i+\frac{1}{2}}k_c}{h_i}\right)u_{i+1}+k_c\left[\frac{r_{i+\frac{1}{2}}}{h_i}+\frac{r_{i-\frac{1}{2}}}{h_{i-1}}\right]u_i-\left(\frac{r_{i-\frac{1}{2}}k_c}{h_{i-1}}\right)u_{i-1}=0,\ i=j+1,\ldots,N$$

with $u_{N+1} = T_c$

Nonlinear Second-Order Equations

We now consider finite difference methods for the solution of nonlinear boundary-value problems of the form

$$Ly(x) \equiv -y'' + f(x, y, y') = 0, \qquad a < x < b \tag{2.89a}$$

$$y(a) = \alpha, \qquad y(b) = \beta \tag{2.89b}$$

If a uniform mesh is used, then a second-order difference approximation to (2.89) is:

$$L_h u_i \equiv -\left[\frac{u_{i+1} - 2u_i + u_{i-1}}{h^2}\right] + f\left(x_i, u_i, \frac{u_{i+1} - u_{i-1}}{2h}\right) = 0,$$

$$i = 1, 2, \ldots, N \tag{2.90}$$

$$u_0 = \alpha, \qquad u_{N+1} = \beta$$

The resulting difference equations (2.90) are in general nonlinear, and we shall use Newton's method to solve them (see Appendix B). We first write (2.90) in the form

$$\mathbf{\Phi}(\mathbf{u}) = \mathbf{0} \tag{2.91}$$

where

$$\mathbf{u} = [u_1, u_2, \ldots, u_N]^T$$

$$\mathbf{\Phi} = [\Phi_1(\mathbf{u}), \Phi_2(\mathbf{u}), \ldots, \Phi_N(\mathbf{u})]^T$$

and

$$\Phi_i(\mathbf{u}) = \frac{h^2}{2} L_h u_i$$

The Jacobian of $\mathbf{\Phi}(\mathbf{u})$ is the tridiagonal matrix

$$J(\mathbf{u}) = \frac{\partial \mathbf{\Phi}(\mathbf{u})}{\partial \mathbf{u}} = \begin{bmatrix} B_1(\mathbf{u}) & C_1(\mathbf{u}) & & & 0 \\ A_2(\mathbf{u}) & B_2(\mathbf{u}) & C_2(\mathbf{u}) & & \\ & \ddots & \ddots & \ddots & \\ & & & B_{N-1}(\mathbf{u}) & C_{N-1}(\mathbf{u}) \\ 0 & & & A_N(\mathbf{u}) & B_N(\mathbf{u}) \end{bmatrix} \tag{2.92}$$

where

$$A_i(\mathbf{u}) = -\frac{1}{2}\left[1 + \frac{h}{2}\frac{\partial f}{\partial y'}\left(x_i, u_i, \frac{u_{i+1} - u_{i-1}}{2h}\right)\right], \qquad i = 2, 3, \ldots, N$$

$$B_i(\mathbf{u}) = \left[1 + \frac{h^2}{2}\frac{\partial f}{\partial y}\left(x_i, u_i, \frac{u_{i+1} - u_{i-1}}{2h}\right)\right], \qquad i = 1, 2, \ldots, N$$

$$C_i(\mathbf{u}) = -\frac{1}{2}\left[1 - \frac{h}{2}\frac{\partial f}{\partial y'}\left(x_i, u_i, \frac{u_{i+1} - u_{i-1}}{2h}\right)\right], \qquad i = 1, 2, \ldots, N-1$$

and

$$\frac{\partial f}{\partial y'}\left(x_i, u_i, \frac{u_{i+1} - u_{i-1}}{2h}\right) \quad \text{is} \quad \frac{\partial f}{\partial y'}(x_i, y, y')$$

with

$$y \quad \text{evaluated by} \quad u_i$$

and

$$y' \quad \text{evaluated by} \quad \frac{u_{i+1} - u_{i-1}}{2h}$$

In computing $\Phi_1(\mathbf{u})$, $\Phi_N(\mathbf{u})$, $A_N(\mathbf{u})$, $B_1(\mathbf{u})$, $B_N(\mathbf{u})$, and $C_1(\mathbf{u})$, we use $u_0 = \alpha$ and $u_{N+1} = \beta$. Now, with any initial estimate $\mathbf{u}^{[0]}$ of the quantities u_i, $i = 1, 2 \ldots, N$, we define

$$\mathbf{u}^{[k+1]} = \mathbf{u}^{[k]} + \Delta\mathbf{u}^{[k]}, \qquad k = 0, 1, 2, \ldots \tag{2.93}$$

where $\Delta\mathbf{u}^{[k]}$ is the solution of

$$J(\mathbf{u}^{[k]})\Delta\mathbf{u}^{[k]} = -\mathbf{\Phi}(\mathbf{u}^{[k]}), \qquad k = 0, 1, 2, \ldots \tag{2.94}$$

More general boundary conditions than those in (2.89b) are easily incorporated into the difference scheme.

EXAMPLE 5

A class of problems concerning diffusion of oxygen into a cell in which an enzyme-catalyzed reaction occurs has been formulated and studied by means of singular perturbation theory by Murray [7]. The transport equation governing the steady concentration C of some substrate in an enzyme-catalyzed reaction has the general form

$$\nabla(D\nabla C) = g(C)$$

Here D is the molecular diffusion coefficient of the substrate in the medium containing uniformly distributed bacteria and $g(C)$ is proportional to the reaction rate. We consider the case with constant diffusion coefficient D_0 in a spherical cell with a Michaelis-Menten theory reaction rate. In dimensionless variables the diffusion kinetics equation can now be written as

$$Ly = (x^2y')' = x^2f(y), \qquad 0 < x < 1$$

where

$$x = \frac{r}{R}, \qquad y(x) = \frac{C(r)}{C_0}, \qquad \varepsilon = \left(\frac{D_0C_0}{nqR^2}\right), \qquad f(y) = \varepsilon^{-1}\frac{y(x)}{y(x) + k}, \qquad k = \frac{k_m}{C_0}$$

Here R is the radius of the cell, C_0 is the constant concentration of the substrate in $r > R$, k_m is the Michaelis constant, q is the maximum rate at which each cell can operate, and n is the number of cells.

Assuming the cell membrane to have infinite permeability, it follows that

$$y(1) = 1$$

Further, from the assumed continuity and symmetry of $y(x)$ with respect to $x = 0$, we must have

$$y'(0) = 0$$

There is no closed-form analytical solution to this problem. Thus, solve this problem using a finite difference method.

SOLUTION

The governing equation is

$$2xy' + x^2y'' = x^2f(y) \quad \text{or} \quad y'' + \frac{2}{x}y' - f(y) = 0$$

with $y(1) = 1$ and $y'(0) = 0$. With the mesh spacing $h = 1/(N + 1)$ and mesh point $x_i = ih$,

$$\left(1 + \frac{1}{i}\right) u_{i+1} - 2u_i + \left(1 - \frac{1}{i}\right) u_{i-1} - h^2 f(u_i) = 0, \qquad i = 1, 2, \ldots, N$$

with $u_{N+1} = 1.0$. For $x = 0$, the second term in the differential equation is evaluated using L'Hospital's rule:

$$\operatorname*{Lim}_{x \to 0} \left(\frac{y'}{x}\right) = \frac{y''}{1}$$

Therefore, the differential equation becomes

$$3y'' - f(y) = 0$$

at $x = 0$, for which the corresponding difference replacement is

$$u_1 - 2u_0 + u_{-1} - \frac{h^2}{3} f(u_0) = 0$$

Using the boundary condition $y'(0) = 0$ gives

$$u_1 - u_0 - \frac{h^2}{6} f(u_0) = 0$$

The vector $\mathbf{\Phi}(\mathbf{u})$ becomes

$$\mathbf{\Phi}(\mathbf{u}) = \begin{bmatrix} u_1 - u_0 - \frac{h^2}{6} f(u_0) \\ \vdots \\ \left(1 + \frac{1}{i}\right) u_{i+1} - 2u_i + \left(1 - \frac{1}{i}\right) u_{i-1} - h^2 f(u_i) \\ \vdots \\ \left(1 + \frac{1}{N}\right) - 2u_N + \left(1 - \frac{1}{N}\right) u_{N-1} - h^2 f(u_N) \end{bmatrix}$$

and the Jacobian is

$$J(\mathbf{u}) = \begin{bmatrix} B_0 & C_0 & & & \\ A_1 & B_1 & C_1 & & \\ & \cdot & \cdot & \cdot & \\ & & \cdot & \cdot & \cdot \\ & & \cdot & \cdot & C_{N-1} \\ & & & A_N & B_N \end{bmatrix}$$

where

$$B_0 = -\left(1 + \frac{h^2}{6}\frac{\partial f}{\partial y}(u_0)\right) = -\left(1 + \frac{h^2}{6\varepsilon} \times \frac{k}{(u_0 + k)^2}\right)$$

$$C_0 = 1$$

$$A_i = \left(1 - \frac{1}{i}\right), \qquad i = 1, 2, \ldots, N$$

$$B_i = -\left(2 + \frac{h^2}{\varepsilon} \times \frac{k}{(u_i + k)^2}\right), \qquad i = 1, 2, \ldots, N$$

$$C_i = \left(1 + \frac{1}{i}\right), \qquad i = 1, 2, \ldots, N - 1$$

Therefore, the matrix equation (2.94) for this problem involves a tridiagonal linear system of order $N + 1$.

The numerical results are shown in Table 2.4. For increasing values of N, the approximate solution appears to be converging to a solution. Decreasing the value of TOL below 10^{-6} gave the same results as shown in Table 2.4; thus the differences in the solutions presented are due to the error in the finite difference approximations. These results are consistent with those presented in Chapter 6 of [1].

The results shown in Table 2.4 are easy to interpret from a physical standpoint. Since y represents the dimensionless concentration of a substrate, and since the substrate is consumed by the cell, the value of y can never be negative and should decrease as x decreases (moving from the surface to the center of the cell).

First-Order Systems

In this section we shall consider the general systems of m first-order equations subject to linear two-point boundary conditions:

$$L\mathbf{y} = \mathbf{y}' - \mathbf{f}(x, \mathbf{y}) = \mathbf{0}, \qquad a < x < b \tag{2.95a}$$

$$A\mathbf{y}(a) + B\mathbf{y}(b) = \boldsymbol{\alpha} \tag{2.95b}$$

TABLE 2.4 Results of Example 5, TOL = (−6) on Newton Iteration, ε = 0.1, k = 0.1

x	$N = 5$	$N = 10$	$N = 20$	$N = 40$	$N = 80$
0.0	0.283(−1)	0.243(−1)	0.232(−1)	0.229(−1)	0.228(−1)
0.2	0.430(−1)	0.384(−1)	0.372(−1)	0.369(−1)	0.368(−1)
0.4	0.103	0.998(−1)	0.989(−1)	0.987(−1)	0.987(−1)
0.6	0.259	0.257	0.257	0.257	0.257
0.8	0.553	0.552	0.552	0.552	0.552
1.0	1.000	1.000	1.000	1.000	1.000

As before, we take the mesh points on $[a, b]$ as

$$x_i = a + ih, \qquad i = 0, 1, \ldots, N + 1 \qquad (2.96)$$
$$h = \frac{b - a}{N + 1}$$

Let the m-dimensional vector $\mathbf{u}_i$ approximate the solution $y(x_i)$, and approximate (2.95a) by the system of difference equations

$$L_h\mathbf{u}_i = \frac{\mathbf{u}_i - \mathbf{u}_{i-1}}{h} - \mathbf{f}\left(x_{i-1/2}, \frac{\mathbf{u}_i + \mathbf{u}_{i-1}}{2}\right) = \mathbf{0},$$
$$i = 1, 2, \ldots, N + 1 \qquad (2.97a)$$

The boundary conditions are given by

$$A\mathbf{u}_0 + B\mathbf{u}_{N+1} - \boldsymbol{\alpha} = \mathbf{0} \qquad (2.97b)$$

The scheme (2.97) is known as the centered-difference method. The nonlinear term in (2.95a) might have been chosen as

$$\tfrac{1}{2}\left[\mathbf{f}(x_i, \mathbf{u}_i) + \mathbf{f}(x_{i-1}, \mathbf{u}_{i-1})\right] \qquad (2.98)$$

resulting in the trapezoidal rule.

On defining the $m(N + 2)$-dimensional vector $\mathbf{U}$ by

$$\mathbf{U} = [\mathbf{u}_0, \mathbf{u}_1, \ldots, \mathbf{u}_{N+1}]^T \qquad (2.99)$$

(2.97) can be written as the system of $m(N + 2)$ equations

$$\boldsymbol{\Phi}(\mathbf{U}) = \begin{bmatrix} A\mathbf{u}_0 + B\mathbf{u}_{N+1} - \boldsymbol{\alpha} \\ hL_h\mathbf{u}_1 \\ \cdot \\ \cdot \\ \cdot \\ \cdot \\ hL_h\mathbf{u}_{N+1} \end{bmatrix} = \mathbf{0} \qquad (2.100)$$

With some initial guess, $\mathbf{U}^{[0]}$, we now compute the sequence of $\mathbf{U}^{[k]}$'s by

$$\mathbf{U}^{[k+1]} = \mathbf{U}^{[k]} + \Delta\mathbf{U}^{[k]}, \qquad k = 0, 1, 2, \ldots \qquad (2.101)$$

where $\Delta\mathbf{U}^{[k]}$ is the solution of the linear algebraic system

$$\frac{\partial\boldsymbol{\Phi}(\mathbf{U}^{[k]})}{\partial\mathbf{U}}\Delta\mathbf{U}^{[k]} = -\boldsymbol{\Phi}(\mathbf{U}^{[k]}) \qquad (2.102)$$

One of the advantages of writing a BVP as a first-order system is that variable mesh spacings can be used easily. Let

$$a = x_0 < x_1 < \ldots < x_{N+1} = b \qquad (2.103)$$
$$h_i = x_{i+1} - x_i, \qquad h = \max_i h_i$$

be a general partition of the interval $[a, b]$.

The approximation for (2.95) using (2.103) with the trapezoidal rule is

$$\Phi(\mathbf{U}) = \begin{bmatrix} A\mathbf{u}_0 + B\mathbf{u}_{N+1} - \boldsymbol{\alpha} \\ h_0 L_h \mathbf{u}_1 \\ \vdots \\ h_N L_h \mathbf{u}_{N+1} \end{bmatrix} = \mathbf{0} \tag{2.104}$$

where

$$h_{i-1} L_h \mathbf{u}_i = \mathbf{u}_i - \mathbf{u}_{i-1} - \frac{h_{i-1}}{2} [\mathbf{f}(x_i, \mathbf{u}_i) + \mathbf{f}(x_{i-1}, \mathbf{u}_{i-1})], \; i = 1, \ldots, N + 1$$

By allowing the mesh to be graded in the region of a sharp gradient, nonuniform meshes can be helpful in solving problems that possess solutions or derivatives that have sharp gradients.

Higher-Order Methods

The difference scheme (2.63) yields an approximation to the solution of (2.52) with an error that is $0(h^2)$. We shall briefly examine two ways in which, with additional calculations, difference schemes may yield higher-order approximations. These error-reduction procedures are Richardson's extrapolation and deferred corrections.

The basis of Richardson's extrapolation is that the error E_i, which is the difference between the approximation and the true solution, can be written as

$$E_i = h^2 a_1(x_i) + h^4 a_2(x_i) + \ldots \tag{2.105}$$

where the functions $a_j(x_i)$ are independent of h. To implement the method, one solves the BVP using successively smaller mesh sizes such that the larger meshes are subsets of the finer ones. For example, solve the BVP twice, with mesh sizes of h and $h/2$. Let the respective solutions be denoted $u_i(h)$ and $u_i(h/2)$. For any point common to both meshes, $x_i = ih = 2i(h/2)$,

$$y(x_i) - u_i(h) = h^2 a_1(x_i) + h^4 a_2(x_i) + \ldots \tag{2.106}$$

$$y(x_i) - u_i\left(\frac{h}{2}\right) = \frac{h^2}{4} a_1(x_i) + \frac{h^4}{16} a_2(x_i) + \ldots$$

Eliminate $a_1(x_i)$ from (2.106) to give

$$y(x_i) = \frac{4u_i\left(\frac{h}{2}\right) - u_i(h)}{3} + 0(h^4) \tag{2.107}$$

Thus an $0(h^4)$ approximation to $y(x)$ on the mesh with spacing h is given by

$$\bar{u}_i = \frac{4}{3} u_i\left(\frac{h}{2}\right) - \frac{1}{3} u_i(h), \qquad i = 0, 1, \ldots, N + 1 \tag{2.108}$$

A further mesh subdivision can be used to produce a solution with error $0(h^6)$, and so on.

For some problems Richardson's extrapolation is useful, but in general, the method of deferred corrections, which is described next, has proven to be somewhat superior [8].

The method of deferred corrections was introduced by Fox [9], and has since been modified and extended by Pereyra [10–12]. Here, we will outline Pereyra's method since it is used in software described in the next section.

Pereyra requires the BVP to be in the following form:

$$\begin{aligned} \mathbf{y}' &= \mathbf{f}(x, \mathbf{y}), \qquad a < x < b \\ \mathbf{g}(\mathbf{y}(a), \mathbf{y}(b)) &= \mathbf{0} \end{aligned} \tag{2.109}$$

and uses the trapezoidal rule approximation

$$\mathbf{u}_{i+1} - \mathbf{u}_i - \tfrac{1}{2} h \left[\mathbf{f}(x_i, \mathbf{u}_i) + \mathbf{f}(x_{i+1}, \mathbf{u}_{i+1})\right] = h\mathbf{T}(\mathbf{u}_{i+1/2}) \tag{2.110}$$

where $\mathbf{T}(\mathbf{u}_{i+1/2})$ is the truncation error. Next, Pereyra writes the truncation error in terms of higher-order derivatives

$$\mathbf{T}(\mathbf{u}_{i+1/2}) = -\sum_{s=1}^{q} \left[\alpha_s h^{2s} \mathbf{f}_{i+1/2}^{(2s)}\right] + 0(h^{2s+2}) \tag{2.111}$$

where

$$\mathbf{f}_{i+1/2}^{(2s)} = \frac{d^{2s}}{dx^{2s}} \mathbf{f}(x_{i+1/2}, \mathbf{u}_{i+1/2})$$

$$\alpha_s = \frac{s}{2^{2s-1}(2s+1)(2s!)}$$

q = number of terms in series (sets the desired accuracy)

The first approximation $\mathbf{u}_i^{[1]}$ is obtained by solving

$$\begin{aligned} &\mathbf{u}_{i+1}^{[1]} - \mathbf{u}_i^{[1]} - \tfrac{1}{2} h[\mathbf{f}(x_i, \mathbf{u}_i^{[1]}) + \mathbf{f}(x_{i+1}, \mathbf{u}_{i+1}^{[1]})] = \mathbf{0} \\ &i = 0, 1, \ldots, N \\ &\mathbf{g}(\mathbf{u}_0^{[1]}, \mathbf{u}_{N+1}^{[1]}) = \mathbf{0} \end{aligned} \tag{2.112}$$

where the truncation error is ignored. This approximation differs from the true solution by $0(h^2)$.

The process proceeds as follows. An approximate solution $\mathbf{u}^{[k]}$ [differs from the true solution by terms of order $0(h^{2k})$] can be obtained from:

$$\begin{aligned} &\mathbf{u}_{i+1}^{[k]} - \mathbf{u}_i^{[k]} - \tfrac{1}{2} h \left[\mathbf{f}(x_i, \mathbf{u}_i^{[k]}) + \mathbf{f}(x_{i+1}, \mathbf{u}_{i+1}^{[k]})\right] = h\mathbf{T}^{[k-1]}(\mathbf{u}_{i+1/2}^{[k-1]}) \\ &i = 0, 1, \ldots, N \\ &\mathbf{g}(\mathbf{u}_0^{[k]}, \mathbf{u}_{N+1}^{[k]}) = \mathbf{0}) \end{aligned} \tag{2.113}$$

where

$$\mathbf{T}^{[k-1]} = \mathbf{T} \quad \text{with} \quad q = k - 1$$

In each step of (2.113), the nonlinear algebraic equations are solved by Newton's method with a convergence tolerance of less than $0(h^{2k})$. Therefore, using (2.112) gives $\mathbf{u}_i^{[1]}$ $(0(h^2))$, which can be used in (2.113) to give $\mathbf{u}_i^{[2]}$ $(0(h^4))$. Successive iterations of (2.113) with increasing k can give even higher-order accurate approximations.

MATHEMATICAL SOFTWARE

The available software that is based on the methods of this chapter is not as extensive as in the case of IVPs. A subroutine for solving a BVP will be designed in a manner similar to that outlined for IVPs in Chapter 1 except for the fact that the routines are much more specialized because of the complexity of solving BVPs. The software discussed below requires the BVPs to be posed as first-order systems (usually allows for simpler algorithms). A typical calling sequence could be

```
CALL DRIVE (FUNC, DFUNC, BOUND, A, B, U, TOL)
```

where

FUNC = user-written subroutine for evaluating $\mathbf{f}(x, \mathbf{y})$

DFUNC = user-written subroutine for evaluating the Jacobian of $\mathbf{f}(x, \mathbf{y})$

BOUND = user-written subroutine for evaluating the boundary conditions and, if necessary, the Jacobian of the boundary conditions

A = left boundary point

B = right boundary point

U = on input contains initial guess of solution vector, and on output contains the approximate solution

TOL = an error tolerance

This is a simplified calling sequence, and more elaborate ones are actually used in commercial routines.

The subroutine DRIVE must contain algorithms that:

1. Implement the numerical technique
2. Adapt the mesh-size (or redistribute the mesh spacing in the case of non-uniform meshes)
3. Calculate the error so to implement step (2) such that the error does not surpass TOL

Implicit within these steps are the subtleties involved in executing the various techniques, e.g., the position of the orthonormalization points when using superposition.

Each of the major mathematical software libraries—IMSL, NAG, and HARWELL—contains routines for solving BVPs. IMSL contains a shooting routine and a modified version of DD04AD (to be described below) that uses a variable-order finite difference method combined with deferred corrections. HARWELL possesses a multiple shooting code and DD04AD. The NAG library includes various shooting codes and also contains a modified version of DD04AD. Software other than that of IMSL, HARWELL, and NAG that is available is listed in Table 2.5. From this table and the routines given in the main libraries, one can see that the software for solving BVPs uses the techniques that are outlined in this chapter.

We illustrate the use of BVP software packages by solving a fluid mechanics problem. The following codes were used in this study:

1. HARWELL, DD03AD (multiple shooting)
2. HARWELL, DD04AD

Notice we have chosen a shooting and a finite difference code. The third major method, superposition, was not used in this study. The example problem is nonlinear and would thus require the use of SUPORQ if superposition is to be included in this study. At the time of this writing SUPORQ is difficult to implement and requires intimate knowledge of the code for effective utilization. Therefore, it was excluded from this study. DD03AD and DD04AD will now be described in more detail.

DD03AD [18]

This program is the multiple-shooting code of the Harwell library. In this algorithm, the interval is subdivided and "shooting" occurs in both directions. The boundary-value problem must be specified as an initial-value problem with the code or the user supplying the initial conditions. Also, the partitioning of the interval can be user-supplied or performed by the code. A tolerance parameter (TOL) controls the accuracy in meeting the continuity conditions at the

TABLE 2.5 BVP Codes

Name	Method Implemented	Reference
BOUNDS	Multiple shooting	[13,14]
SHOOT1	Shooting with separated boundary conditions	[15]
SHOOT2	Same as SHOOT1 with more general boundary conditions	[15]
MSHOOT	Mutliple shooting	[15]
SUPORT	Superposition (linear problems only)	[4]
SUPORQ	Superposition with quasilinearization	[16]

matching points [see (2.35)]. This type of code takes advantage of the highly developed software available for IVPs (uses a fourth-order Runge-Kutta algorithm [19]).

DD04AD [17, 20]

This code was written by Lentini and Pereyra and is described in detail in [20]. Also, an earlier version of the code is discussed in [17]. The code implements the trapezoidal approximation, and the resulting algebraic system is solved by a modified Newton method. The user is permitted to specify an initial interval partition (which does not need to be uniform), or the code provides a coarse, equispaced one. The user may also specify an initial estimate for the solution (the default being zero). Deferred corrections is used to increase accuracy and to calculate error estimates. An error tolerance (TOL) is provided by the user. Additional mesh points are automatically added to the initial partition with the aim of reducing error to the user-specified level, and also with the aim of equidistribution of error throughout the interval [17]. The new mesh points are always added between the existing mesh points. For example, if x_j and x_{j+1} are initial mesh points, then if m mesh points t_i, $i = 1, 2, \ldots, m$, are required to be inserted into $[x_j, x_{j+1}]$, they are placed such that

$$x_j < t_1 < \ldots < t_m < x_{j+1} \tag{2.114}$$

where

$$t_1 = \frac{t_2 - x_j}{2}, \ldots, t_i = \frac{t_{i+1} - t_{i-1}}{2}, \ldots, t_m = \frac{x_{j+1} - t_{m-1}}{2}$$

The approximate solution is given as a discrete function at the points of the final mesh.

Example Problem

The following BVP arises in the study of the behavior of a thin sheet of viscous liquid emerging from a slot at the base of a converging channel in connection with a method of lacquer application known as "curtain coating" [21]:

$$\frac{d^2y}{dx^2} - \frac{1}{y}\left(\frac{dy}{dx}\right)^2 - y\frac{dy}{dx} + 1 = 0 \tag{2.115}$$

The function y is the dimensionless velocity of the sheet, and x is the dimensionless distance from the slot. Appropriate boundary conditions are [22]:

$$y = y_0 \quad \text{at} \quad x = 0 \tag{2.116}$$

$$\frac{dy}{dx} \rightarrow (2x)^{-1/2} \quad \text{at} \quad \text{sufficiently large } x$$

In [22] (2.115) was solved using a reverse shooting procedure subject to the boundary conditions

$$y = 0.325 \quad \text{at} \quad x = 0 \tag{2.117}$$

and

$$\frac{dy}{dx} = (2x)^{-1/2} \quad \text{at} \quad x = x_R$$

The choice of $x_R = 50$ was found by experimentation to be optimum in the sense that it was large enough for (2.116) to be "sufficiently valid." For smaller values of x_R, the values of y at zero were found to have a variation of as much as 8%.

We now study this problem using DD03AD and DD04AD. The results are shown in Table 2.6. DD03AD produced approximate solutions only when a large number of shooting points were employed. Decreasing TOL from 10^{-4} to 10^{-6} when using DD03AD did not affect the results, but increasing the number of shooting points resulted in drastic changes in the solution. Notice that the boundary condition at $x = 0$ is never met when using DD03AD, even when using a large number of shooting points (SP = 360). Davis and Fairweather [23] studied this problem, and their results are shown in Table 2.6 for comparison. DD04AD was able to produce the same results as Davis and Fairweather in significantly less execution time than DD03AD.

We have surveyed the types of BVP software but have not attempted to make any sophisticated comparisons. This is because in the author's opinion, based upon the work already carried out on IVP solvers, there is no sensible basis for comparing BVP software.

Like IVP software, BVP codes are not infallible. If you obtain spurious results from a BVP code, you should be able to rationalize your data with the aid of the code's documentation and the material presented in this chapter.

TABLE 2.6 Results of Eq. (2.115) with (2.117) and $x_R = 5.0$

x	DD03AD TOL = 10^{-4}, SP = 80†	DD03AD TOL = 10^{-6}, SP = 80	DD03AD TOL = 10^{-6}, SP = 320	DD04AD TOL = 10^{-4}	Reference [23] TOL = 10^{-4}
0.0	0.3071	0.3071	0.3205	0.3250	0.3250
1.0	0.9115	0.9115	0.9253	0.9299	0.9299
2.0	0.1462(1)	0.1462(1)	0.1474(1)	0.1477(1)	0.1477(1)
3.0	0.1931(1)	0.1931(1)	0.1941(1)	0.1945(1)	0.1945(1)
4.0	0.2340(1)	0.2340(1)	0.2349(1)	0.2349(1)	0.2349(1)
5.0	0.2737(1)	0.2737(1)	0.2743(1)	0.2701(1)	0.2701(1)
E.T.R.‡	3.75	4.09	14.86	1.0	—

† SP = number of "shooting" points.

‡ E.T.R. = Execution time ratio = $\dfrac{\text{execution time}}{\text{execution time of DD04AD with TOL} = 10^{-4}}$.

PROBLEMS

1. Consider the BVP

$$y'' + r(x)y = f(x), \qquad a < x < b$$
$$y(a) = \alpha$$
$$y(b) = \beta$$

Show that for a uniform mesh, the integration method gives the same result as Eq. (2.64).

2. Refer to Example 4. If

$$S(r) = S_0\left[1 + b\left(\frac{r}{r_f}\right)^2\right]$$

solve the governing differential equation to obtain the temperature profile in the core and the cladding. Compare your results with the analytical solution given on page 304 of [24]. Let $k_c = 0.64$ cal/(s·cm·K), $k_f = 0.066$ cal/(s·cm·K), $T_0 = 500$ K, $r_c = \frac{1}{2}$ in, and $r_f = \frac{3}{8}$ in.

3.* Axial conduction and diffusion in an adiabatic tubular reactor can be described by [2]:

$$\frac{1}{\text{Pe}}\frac{d^2C}{dx^2} - \frac{dC}{dx} - R(C, T) = 0$$

$$\frac{1}{\text{Bo}}\frac{d^2T}{dx^2} - \frac{dT}{dx} - \beta R(C, T) = 0$$

with

$$\left.\begin{aligned}\frac{1}{\text{Pe}}\frac{dC}{dx} &= C - 1\\ \frac{1}{\text{Bo}}\frac{dT}{dx} &= T - 1\end{aligned}\right\} \quad \text{at} \quad x = 0$$

and

$$\left.\begin{aligned}\frac{dC}{dx} &= 0\\ \frac{dT}{dx} &= 0\end{aligned}\right\} \quad \text{at} \quad x = 0$$

Calculate the dimensionless concentration C and temperature T profiles for $\beta = -0.05$, Pe = Bo = 10, $E = 18$, and $R(C, T) = 4C \exp[E(1 - 1/T)]$.

4.* Refer to Example 5. In many reactions the diffusion coefficient is a function of the substrate concentration. The diffusion coefficient can be of form [1]:

$$\frac{D(y)}{D_0} = 1 + \frac{\lambda}{(y + k_2)^2}$$

Computations are of interest for $\lambda = k_2 = 10^{-2}$, with ε and k as in Example 5. Solve the transport equation using $D(y)$ instead of D for the parameter choice stated above. Next, let $\lambda = 0$ and show that your results compare with those of Table 2.4.

5.* The reactivity behavior of porous catalyst particles subject to both internal mass concentration gradients as well as temperature gradients can be studied with the aid of the following material and energy balances:

$$\frac{d^2y}{dx^2} + \frac{2}{x}\frac{dy}{dx} = \phi^2 y \exp\left[\gamma\left(1 - \frac{1}{T}\right)\right]$$

$$\frac{d^2T}{dx^2} + \frac{2}{x}\frac{dT}{dx} = -\beta\phi^2 y \exp\left[\gamma\left(1 - \frac{1}{T}\right)\right]$$

with

$$\frac{dT}{dx} = \frac{dy}{dx} = 0 \quad \text{at} \quad x = 0$$

$$T = y = 1 \quad \text{at} \quad x = 1$$

where

y = dimensionless concentration
T = dimensionless temperature
x = dimensionless radial coordiante (spherical geometry)
ϕ = Thiele modulus (first-order reaction rate)
γ = Arrhenius number
β = Prater number

These equations can be combined into a single equation such that

$$\frac{d^2y}{dx^2} + \frac{2}{x}\frac{dy}{dx} = \phi^2 y \exp\left[\frac{\gamma\beta(1 - y)}{1 + \beta(1 - y)}\right]$$

with

$$\frac{dy}{dx} = 0 \quad \text{at} \quad x = 0$$

$$y = 1 \quad \text{at} \quad x = 1$$

For $\gamma = 30$, $\beta = 0.4$, and $\phi = 0.3$, Weisz and Hicks [25] found three solutions to the above equation using a shooting method. Calculate the dimensionless concentration and temperature profiles of the three solutions.

Hint: **Try various initial guesses.**

REFERENCES

1. Keller, H. B., *Numerical Methods for Two-Point Boundary-Value Problems,* Blaisdell, New York (1968).
2. Carberry, J. J., *Chemical and Catalytic Reaction Engineering,* McGraw-Hill, New York (1976).
3. Deuflhard, P., "Recent Advances in Multiple Shooting Techniques," in *Computational Techniques for Ordinary Differential Equations,* I. Gladwell and D. K. Sayers (eds.), Academic, London (1980).
4. Scott, M. R., and H. A. Watts, SUPORT—A Computer Code for Two-Point Boundary-Value Problems via Orthonormalization, SAND75-0198, Sandia Laboratories, Albuquerque, N. Mex. (1975).
5. Scott, M. R., and H. A. Watts, "Computational Solutions of Linear Two-Point Boundary Value Problems via Orthonormalization," SIAM J. Numer. Anal., *14,* 40 (1977).
6. Varga, R. S., *Matrix Iterative Analysis,* Prentice-Hall, Englewood Cliffs, N.J. (1962).
7. Murray, J. D., "A Simple Method for Obtaining Approximate Solutions for a Large Class of Diffusion-Kinetic Enzyme Problems," Math. Biosci., *2* (1968).
8. Fox, L., "Numerical Methods for Boundary-Value Problems," in *Computational Techniques for Ordinary Differential Equations,* I. Gladwell and D. K. Sayers (eds.), Academic, London (1980).
9. Fox, L., "Some Improvements in the Use of Relaxation Methods for the Solution of Ordinary and Partial Differential Equations," Proc. R. Soc. A, *190,* 31 (1947).
10. Pereyra, V., "The Difference Correction Method for Non-Linear Two-Point Boundary Problems of Class M," Rev. Union Mat. Argent. *22,* 184 (1965).
11. Pereyra, V., "High Order Finite Difference Solution of Differential Equations," STAN-CS-73-348, Computer Science Dept., Stanford Univ., Stanford, Calif. (1973).
12. Keller, H. B., and V. Pereyra, "Difference Methods and Deferred Corrections for Ordinary Boundary Value Problems," SIAM J. Numer. Anal., *16,* 241 (1979).

13. Bulirsch, R., "Multiple Shooting Codes," in *Codes for Boundary-Value Problems in Ordinary Differential Equations,* Lecture Notes in Computer Science, *76,* Springer-Verlag, Berlin (1979).

14. Deuflhard, P., "Nonlinear Equation Solvers in Boundary-Value Codes," Rep. TUM-MATH-7812. Institut fur Mathematik, Universitat Munchen (1979).

15. Scott, M. R. and H. A. Watts, "A Systematized Collection of Codes for Solving Two-Point Boundary-Value Problems," *Numerical Methods for Differential Systems,* L. Lapidus and W. E. Schiesser (eds.), Academic, New York (1976).

16. Scott, M. R., and H. A. Watts, "Computational Solution of Nonlinear Two-Point Boundary Value Problems," Rep. SAND 77-0091, Sandia Laboratories, Albuquerque, N. Mex. (1977).

17. Lentini, M., and V. Pereyra, "An Adaptive Finite Difference Solver for Nonlinear Two-Point Boundary Problems with Mild Boundary Layers," SIAM J. Numer. Anal., *14,* 91 (1977).

18. England, R., "A Program for the Solution of Boundary Value Problems for Systems of Ordinary Differential Equations," Culham Lab., Abingdon: Tech. Rep. CLM-PDM 3/73 (1976).

19. England, R., "Error Estimates for Runge-Kutta Type Solutions to Systems of Ordinary Differential Equations," Comput. J., *12,* 166 (1969).

20. Pereyra, V., "PASVA3—An Adaptive Finite-Difference FORTRAN Program for First-Order Nonlinear Ordinary Boundary Problems," in *Codes for Boundary-Value Problems in Ordinary Differential Equations,* Lecture Notes in Computer Science, *76,* Springer-Verlag, Berlin (1979).

21. Brown, D. R., "A Study of the Behavior of a Thin Sheet of Moving Liquid," J. Fluid Mech., *10,* 297 (1961).

22. Salariya, A. K., "Numerical Solution of a Differential Equation in Fluid Mechanics," Comput. Methods Appl. Mech. Eng., *21,* 211 (1980).

23. Davis, M., and G. Fairweather, "On the Use of Spline Collocation for Boundary Value Problems Arising in Chemical Engineering," Comput. Methods Appl. Mech. Eng., *28,* 179 (1981).

24. Bird, R. B., W. E. Stewart, and E. L. Lightfoot, *Transport Phenomena,* Wiley, New York (1960).

25. Weisz, P. B., and J. S. Hicks, "The Behavior of Porous Catalyst Particles in View of Internal Mass and Heat Diffusion Effects," Chem. Eng. Sci., *17,* 265 (1962).

BIBLIOGRAPHY

For additional or more detailed information concerning boundary-value problems, see the following:

Aziz, A. Z. (ed.), *Numerical Solutions of Boundary-Value Problems for Ordinary Differential Equations,* Academic, New York (1975).

Childs, B., M. Scott, J. W. Daniel, E. Denman, and P. Nelson (eds.), *Codes for Boundary-Value Problems in Ordinary Differential Equations,* Lecture Notes in Computer Science, Volume 76, Springer-Verlag, Berlin (1979).

Fox, L., *The Numerical Solution of Two-Point Boundary-Value Problems in Ordinary Differential Equations,* (1957).

Gladwell, I., and D. K. Sayers (eds.), *Computational Techniques for Ordinary Differential Equations,* Academic, London (1980).

Isaacson, E., and H. B. Keller, *Analysis of Numerical Methods,* Wiley, New York (1966).

Keller, H. B., *Numerical Methods for Two-Point Boundary-Value Problems,* Blaisdell, New York (1968).

Russell, R. D., *Numerical Solution of Boundary Value Problems,* Lecture Notes, Universidad Central de Venezuela, Publication 79-06, Caracas (1979).

Varga, R. S., *Matrix Iterative Analysis,* Prentice-Hall, Englewood Cliffs, N.J. (1962).

3

Boundary-Value Problems for Ordinary Differential Equations: Finite Element Methods

INTRODUCTION

The numerical techniques outlined in this chapter produce approximate solutions that, in contrast to those produced by finite difference methods, are continuous over the interval. The approximate solutions are piecewise polynomials, thus qualifying the techniques to be classified as finite element methods [1]. Here, we discuss two types of finite element methods: collocation and Galerkin.

BACKGROUND

Let us begin by illustrating finite elements methods with the following BVP:

$$y'' = y + f(x), \qquad 0 < x < 1 \tag{3.1a}$$

$$\begin{aligned} y(0) &= 0 \\ y(1) &= 0 \end{aligned} \tag{3.1b}$$

Finite element methods find a piecewise polynomial (pp) approximation, $u(x)$, to the solution of (3.1). A piecewise polynomial is a function defined on a partition such that on the subintervals defined by the partition, it is a polynomial. The pp-approximation can be represented by

$$u(x) = \sum_{j=1}^{m} a_j \phi_j(x) \tag{3.2}$$

where $\{\phi_j(x)|j=1, \ldots, m\}$ are specified functions that are piecewise continuously differentiable, and $\{a_j|j=1, \ldots, m\}$ are as yet unknown constants. For now, assume that the functions $\phi_j(x)$, henceforth called basis functions (to be explained in the next section), satisfy the boundary conditions. The finite element methods differ only in how the unknown coefficients $\{a_j|j=1, \ldots, m\}$ are determined.

In the collocation method, the set $\{a_j|j=1, \ldots, m\}$ is determined by satisfying the BVP exactly at m points, $\{x_i|i=1, \ldots, m\}$, the collocation points, in the interval. For (3.1):

$$u''(x_i) - u(x_i) - f(x_i) = 0, \qquad i = 1, \ldots, m \tag{3.3}$$

If $u(x)$ is given by (3.2), then (3.3) becomes

$$\sum_{j=1}^{m} a_j[\phi_j''(x_i) - \phi_j(x_i)] - f(x_i) = 0, \qquad i = 1, \ldots, m \tag{3.4}$$

or in matrix notation,

$$A^C\mathbf{a} = \mathbf{f} \tag{3.5}$$

where

$$A_{ij}^C = \phi_j''(x_i) - \phi_j(x_i)$$

$$\mathbf{a} = [a_1, a_2, \ldots, a_m]^T$$

$$\mathbf{f} = [f(x_1), f(x_2), \ldots, f(x_m)]^T$$

The solution of (3.5) then yields the vector $\mathbf{a}$, which determines the collocation approximation (3.2).

To formulate the Galerkin method, first multiply (3.1) by ϕ_i and integrate the resulting equation over [0, 1]:

$$\int_0^1 [y''(x) - y(x) - f(x)]\phi_i(x)\, dx = 0, \qquad i = 1, \ldots, m \tag{3.6}$$

Integration of $y''(x)\phi_i(x)$ by parts gives

$$\int_0^1 y''(x)\phi_i(x)\, dx = y'(x)\phi_i(x)\Big|_0^1 - \int_0^1 y'(x)\phi_i'(x)\, dx, \qquad i = 1, \ldots, m$$

Since the functions $\phi_i(x)$ satisfy the boundary conditions, (3.6) becomes

$$\int_0^1 y'(x)\phi_i'(x)\, dx + \int_0^1 [y(x) + f(x)]\phi_i(x)\, dx = 0, \qquad i = 1, \ldots, m \tag{3.7}$$

For any two functions η and ψ we define

$$(\eta, \psi) = \int_0^1 \eta(x)\psi(x)\, dx \tag{3.8}$$

With (3.8), Eq. (3.7) becomes

$$(y', \phi_i') + (y, \phi_i) + (f, \phi_i) = 0, \qquad i = 1, \ldots, m \tag{3.9}$$

and is called the weak form of (3.1). The Galerkin method consists of finding $u(x)$ such that

$$(u', \phi_i') + (u, \phi_i) + (f, \phi_i) = 0, \qquad i = 1, \ldots, m \tag{3.10}$$

If $u(x)$ is given by (3.2), then (3.10) becomes:

$$\left(\sum_{j=1}^{m} a_j\phi_j', \phi_i'\right) + \left(\sum_{j=1}^{m} a_j\phi_j, \phi_i\right) + (f, \phi_i) = 0, \qquad i = 1, \ldots, m \tag{3.11}$$

or, in matrix notation,

$$A^G\mathbf{a} = -\mathbf{g} \tag{3.12}$$

where

$$A_{ij}^G = (\phi_j', \phi_i') + (\phi_j, \phi_i)$$

$$\mathbf{a} = [a_1, \ldots, a_m]^T$$

$$\mathbf{g} = [\bar{f}_1, \ldots, \bar{f}_m]^T$$

$$\bar{f}_i = (f, \phi_i)$$

The solution of (3.12) gives the vector **a**, which specifies the Galerkin approximation (3.2).

Before discussing these methods in further detail, we consider choices of the basis functions.

PIECEWISE POLYNOMIAL FUNCTIONS

To begin the discussion of pp-functions, let the interval partition π be given by:

$$a = x_1 < x_2 < \ldots < x_{\ell+1} = b \tag{3.13}$$

with

$$h = \max_{1 \le j \le \ell} h_j = \max_{1 \le j \le \ell} (x_{j+1} - x_j)$$

Also let $\{P_j(x) | j = 1, \ldots, \ell\}$ be any sequence of ℓ polynomials of order k (degree $\le k - 1$). The corresponding pp-function, $F(x)$, of order k is defined by

$$F(x) = P_j(x), \qquad x_j < x < x_{j+1} \tag{3.14}$$

$$j = 1, \ldots, \ell$$

where x_j are called the breakpoints of F. By convention

$$F(x) = \begin{cases} P_1(x), & x \leqslant x_1 \\ P_\ell(x), & x \geqslant x_{\ell+1} \end{cases} \tag{3.15}$$

and

$$F(x_i) = P_i(x_i) \qquad \text{(right continuity)} \tag{3.16}$$

A portion of a pp-function is illustrated in Figure 3.1. The problem is how to conveniently represent the pp-function.

Let S be a set of functions:

$$S = \{\lambda_j(x) | j = 1, \ldots, L\} \tag{3.17}$$

The class of functions denoted by $\mathscr{L}$ is defined to be the set of all functions $f(x)$ of the form

$$f(x) = \sum_{j=1}^{L} \alpha_j \lambda_j(x) \tag{3.18}$$

where the α_j's are constants. This class of functions $\mathscr{L}$ defined by (3.18) is called a linear function space. This is analogous to a linear vector space, for if vectors $\mathbf{x}_j$ are substituted for the functions $\lambda_j(x)$ in (3.18), we have the usual definition of an element $\mathbf{x}$ of a vector space. If the functions λ_j in S are linearly independent, then the set S is called a basis for the space $\mathscr{L}$, L is the dimension of the space $\mathscr{L}$, and each function λ_j is called a basis function.

Define $\mathscr{L}_k(\pi)$ (subspace of $\mathscr{L}$) to be the set of all pp-functions of order k on the partition π. The dimension of this space is

$$\dim \mathscr{L}_k(\pi) = k\ell \tag{3.19}$$

Let $\boldsymbol{\nu}$ be a sequence of nonnegative integers ν_j, that is, $\boldsymbol{\nu} = \{\nu_j | j = 2, \ldots, \ell\}$, such that

$$\text{jump}_{x_j} \frac{d^{i-1}}{dx^{i-1}} [f(x)] = 0 \tag{3.20}$$

$$i = 1, \ldots, \nu_j, \qquad j = 2, \ldots, \ell$$

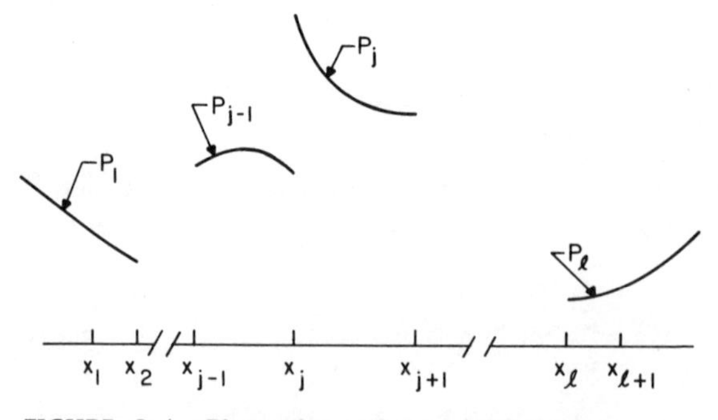

FIGURE 3.1 Piecewise polynomial function.

where

$$\text{jump}_{x_j} \frac{d^{i-1}}{dx^{i-1}} [f(x)] = \frac{d^{i-1}}{dx^{i-1}} [f(x_j^+)] - \frac{d^{i-1}}{dx^{i-1}} [f(x_j^-)] \tag{3.21}$$

or in other words, $\boldsymbol{\nu}$ specifies the continuity (if any) of the function and its derivative at the breakpoints. Define the subspace $\mathscr{P}_k^{\boldsymbol{\nu}}(\pi)$ of $\mathscr{P}_k(\pi)$ by

$$\mathscr{P}_k^{\boldsymbol{\nu}}(\pi) = \left\{ \begin{matrix} f(x) \text{ is in } \mathscr{P}_k(\pi) \text{ and satisfies the jump} \\ \text{conditions specified by } \boldsymbol{\nu} \end{matrix} \right\} \tag{3.22}$$

The dimension of the space $\mathscr{P}_k^{\boldsymbol{\nu}}(\pi)$ is

$$\dim \mathscr{P}_k^{\boldsymbol{\nu}}(\pi) = \sum_{j=1}^{\ell} (k - \nu_j) \tag{3.23}$$

where $\nu_1 = 0$.

We now have a space, $\mathscr{P}_k^{\boldsymbol{\nu}}(\pi)$, that can contain pp-functions such as $F(x)$. Since the λ_i's can be a basis for $\mathscr{P}_k^{\boldsymbol{\nu}}(\pi)$, then $F(x)$ can be represented by (3.18). Next, we illustrate various spaces $\mathscr{P}_k^{\boldsymbol{\nu}}(\pi)$ and bases for these spaces. When using $\mathscr{P}_k^{\boldsymbol{\nu}}(\pi)$ as an approximating space for solving differential equations by finite element methods, we will not use variable continuity throughout the interval. Therefore, notationally replace $\boldsymbol{\nu}$ by ν, where $\{\nu_j = \nu | j = 2, \ldots, \ell\}$.

The simplest space is $\mathscr{P}_2^1(\pi)$, the space of piecewise linear functions. A basis for this space consists of straight-line segments (degree = 1) with discontinuous derivatives at the breakpoints ($\nu = 1$). This basis is given in Table 3.1 and is shown in Figure 3.2*a*. Notice that the dimension of $\mathscr{P}_2^1(\pi)$ is $\ell + 1$ and that there are $\ell + 1$ basis functions given in Table 3.1. Thus, (3.18) can be written as

$$f(x) = \sum_{j=1}^{\ell+1} \alpha_j w_j \tag{3.24}$$

TABLE 3.1 Linear Basis Functions

$$w_1 = \begin{cases} \dfrac{x_2 - x}{x_2 - x_1}, & \text{for } x_1 \leqslant x \leqslant x_2 \\ 0, & \text{for } x \geqslant x_2 \end{cases}$$

$$w_j = \begin{cases} \dfrac{x - x_{j-1}}{x_j - x_{j-1}}, & \text{for } x_{j-1} \leqslant x \leqslant x_j \\ \dfrac{x_{j+1} - x}{x_{j+1} - x_j}, & \text{for } x_j \leqslant x \leqslant x_{j+1} \\ 0, & \text{for } x_{j-1} \leqslant x,\ x \geqslant x_{j+1} \end{cases}$$

$$w_{\ell+1} = \begin{cases} 0, & \text{for } x \leqslant x_\ell \\ \dfrac{x - x_\ell}{x_{\ell+1} - x_\ell}, & \text{for } x_\ell \leqslant x \leqslant x_{\ell+1} \end{cases}$$

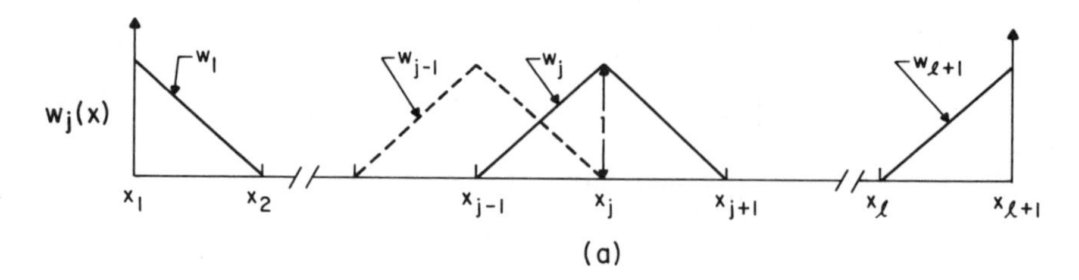

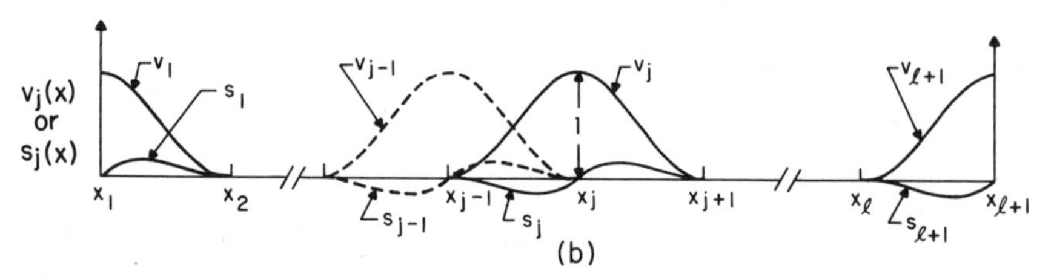

FIGURE 3.2 Schematic of basis functions. (*a*) Piecewise linear functions. (*b*) Piecewise hermite cubic functions.

Frequently, one is interested in numerical approximations that have continuity of derivatives at the interior breakpoints. Obviously, $\mathscr{P}_2^1(\pi)$ does not possess this property, so one must resort to a high-order space.

A space possessing continuity of the first derivative is the Hermite cubic space, $\mathscr{P}_4^2(\pi)$. A basis for this space is the "value" v_j and the "slope" s_j functions given in Table 3.2 and shown in Figure 3.2*b*. Some important properties of this basis are

$$
\begin{aligned}
v_j &= \begin{cases} 0 & \text{at} \quad \text{all } x_i \neq x_j \\ 1 & \text{at} \quad x_j \end{cases} \\
\frac{dv_j}{dx} &= 0 \quad \text{at} \quad \text{all } x_i \\
s_j &= 0 \quad \text{at} \quad \text{all } x_i \\
\frac{ds_j}{dx} &= \begin{cases} 0 & \text{at} \quad \text{all } x_i \neq x_j \\ 1 & \text{at} \quad x_j \end{cases}
\end{aligned}
\tag{3.25}
$$

The dimension of this space is $2(\ell + 1)$; thus (3.18) can be written as

$$
f(x) = \sum_{j=1}^{\ell+1} [\alpha_j^{(1)} v_j + \alpha_j^{(2)} s_j] \tag{3.26}
$$

where $\alpha_j^{(1)}$ and $\alpha_j^{(2)}$ are constants. Since $\nu = 2$, $f(x)$ is continuous and also possesses a continuous first derivative. Notice that because of the properties

TABLE 3.2 Hermite Cubic Basis Functions

$h_j = x_{j+1} - x_j$	$g_1(x) = -2x^3 + 3x^2, \quad 0 \leq x \leq 1$
$\xi_j(x) = \dfrac{x - x_j}{h_j}$	$g_2(x) = x^3 - x^2, \quad 0 \leq x \leq 1$

Value Functions		Slope Functions	
$v_1 = \begin{cases} g_1(1-\xi_1(x)), \\ 0, \end{cases}$	$\begin{matrix} 0 \leq x \leq x_2 \\ x_2 \leq x \leq 1 \end{matrix}$	$s_1 = \begin{cases} -h_1 g_2(1-\xi_1(x)), \\ 0, \end{cases}$	$\begin{matrix} 0 \leq x \leq x_2 \\ x_2 \leq x \leq 1 \end{matrix}$
$v_j = \begin{cases} g_1(\xi_{j-1}(x)), \\ g_1(1-\xi_j(x)), \\ 0, \end{cases}$	$\begin{matrix} x_{j-1} \leq x \leq x_j \\ x_j \leq x \leq x_{j+1} \\ 0 \leq x \leq x_{j-1}, x_{j+1} \leq x \leq 1 \end{matrix}$	$s_j = \begin{cases} h_{j-1} g_2(\xi_{j-1}(x)), \\ -h_j g_2(1-\xi_j(x)), \\ 0, \end{cases}$	$\begin{matrix} x_{j-1} \leq x \leq x_j \\ x_j \leq x \leq x_{j+1} \\ 0 \leq x \leq x_{j-1}, x_{j+1} \leq x \leq 1 \end{matrix}$
$v_{\ell+1} = \begin{cases} 0, \\ g_1(\xi_\ell(x)), \end{cases}$	$\begin{matrix} 0 \leq x \leq x_\ell \\ x_\ell \leq x \leq 1 \end{matrix}$	$s_{\ell+1} = \begin{cases} 0, \\ h_\ell g_2(\xi_\ell(x)), \end{cases}$	$\begin{matrix} 0 \leq x \leq x_\ell \\ x_\ell \leq x \leq 1 \end{matrix}$

shown in (3.25) the vector

$$\boldsymbol{\alpha}^{(1)} = [\alpha_1^{(1)}, \alpha_2^{(1)}, \ldots, \alpha_{\ell+1}^{(1)}]^T$$

give the values of $f(x_i)$, $i = 1, \ldots, \ell + 1$ while the vector

$$\boldsymbol{\alpha}^{(2)} = [\alpha_1^{(2)}, \alpha_2^{(2)}, \ldots, \alpha_{\ell+1}^{(2)}]^T$$

gives the values of $df(x_i)/dx$, $i = 1, \ldots, \ell + 1$. Also, notice that the Hermite cubic as well as the linear basis have limited or local support on the interval; that is, they are nonzero over a small portion of the interval.

A suitable basis for $\mathscr{P}_k^\nu(\pi)$ given any ν, k, and π is the B-spline basis [2]. Since this basis does not have a simple representation like the linear or Hermite cubic basis, we refer the reader to Appendix D for more details on B-splines. Here, we denote the B-spline basis functions by $B_j(x)$ and write (3.18) as:

$$f(x) = \sum_{j=1}^{N} \alpha_j B_j(x) \tag{3.27}$$

where

$$N = \dim \mathscr{P}_k^\nu(\pi)$$

Important properties of the B_j's are:

1. They have local support.
2. $B_1(a) = 1$, $B_N(b) = 1$.
3. Each $B_j(x)$ satisfies $0 \leq B_j(x) \leq 1$ (normalized B-splines).
4. $\sum_{j=1}^{N} B_j(x) = 1$ for $a \leq x \leq b$.

THE GALERKIN METHOD

Consider (3.1) and its weak form (3.9). The use of (3.2) in (3.9) produces the matrix problem (3.12). Since the basis ϕ_i is local, the matrix $\boldsymbol{A}^G$ is sparse.

EXAMPLE 1

Set up the matrix problem for

$$\begin{aligned} -y''(x) &= 1, \qquad 0 < x < 1, \\ y(0) &= 0 \\ y(1) &= 0 \end{aligned}$$

using $\mathscr{P}_2^1(\pi)$ as the approximating space.

SOLUTION

Using $\mathscr{P}_2^1(\pi)$ gives

$$u(x) = \sum_{j=1}^{\ell+1} a_j w_j$$

Since we have imposed the condition that the basis functions satisfy the boundary conditions, the first and last basis function given in Table 3.1 are excluded. Therefore, the pp-approximation is given by

$$u(x) = \sum_{j=1}^{\ell-1} a_j w_j$$

where the w_j's are as shown in Figure 3.3. The matrix $\boldsymbol{A}^G$ is given by

$$A_{ij}^G = \int_0^1 \phi_j' \phi_i' \, dx$$

Because each basis function is supported on only two subintervals [see Figure 3.2(a)],

$$A_{ij}^G = 0 \quad \text{if} \quad |i - j| > 1$$

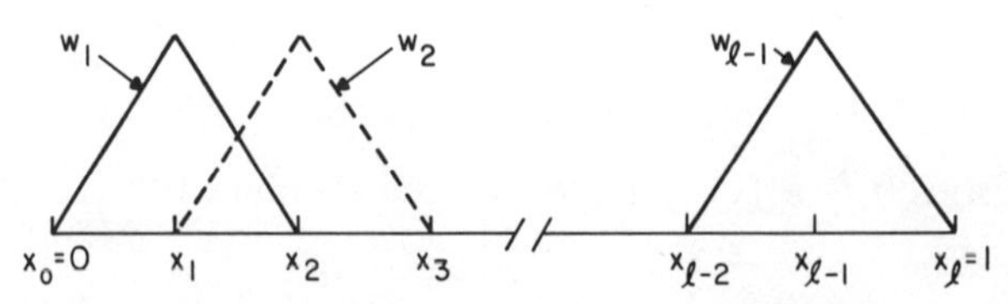

FIGURE 3.3 Numbering of basis functions for Example 1.

Thus, A^G is tridiagonal and

$$A_{ii}^G = \int_0^1 (\phi_i')^2\,dx = \int_{x_{i-1}}^{x_i} \left[\frac{1}{x_i - x_{i-1}}\right]^2 dx + \int_{x_i}^{x_{i+1}} \left[\frac{-1}{x_{i+1} - x_i}\right]^2 dx$$

$$= \frac{1}{h_i} + \frac{1}{h_{i+1}}, \qquad h_i = x_i - x_{i-1}$$

$$A_{i,i+1}^G = \int_0^1 \phi_i' \phi_{i+1}'\,dx = \int_{x_i}^{x_{i+1}} \left[\frac{-1}{x_{i+1} - x_i}\right]\left[\frac{1}{x_{i+1} - x_i}\right] dx$$

$$= -\frac{1}{h_{i+1}}$$

$$A_{i,i-1}^G = -\frac{1}{h_i}$$

The vector **g** is given by

$$\bar{f}_i = \int_0^1 f(x)\phi_i(x)\,dx = \int_0^1 \phi_i(x)\,dx = \tfrac{1}{2}[h_i + h_{i+1}]$$

Therefore, the matrix problem is:

$$\begin{bmatrix} \left(\frac{1}{h_1} + \frac{1}{h_2}\right) & -\frac{1}{h_2} & & & 0 \\ -\frac{1}{h_2} & \left(\frac{1}{h_2} + \frac{1}{h_3}\right) & -\frac{1}{h_3} & & \\ \cdot & \cdot & \cdot & & \\ & \cdot & \cdot & \cdot & -\frac{1}{h_{\ell-2}} \\ 0 & & & -\frac{1}{h_{\ell-2}} & \left(\frac{1}{h_{\ell-1}}\right) \end{bmatrix} \begin{bmatrix} a_1 \\ a_2 \\ \cdot \\ a_{\ell-2} \\ a_{\ell-1} \end{bmatrix} = \begin{bmatrix} \frac{1}{2}(h_1 + h_2) \\ \frac{1}{2}(h_2 + h_3) \\ \cdot \\ \cdot \\ \frac{1}{2}(h_{\ell-2} + h_{\ell-1}) \end{bmatrix}$$

From Example 1, one can see that if a uniform mesh is specified using $\mathscr{P}_2^1(\pi)$, the standard second-order correct finite difference method is obtained. Therefore, the method would be second-order accurate. In general, the Galerkin method using $\mathscr{P}_k^\nu(\pi)$ gives an error such that [1]:

$$\|y - u\| \leq Ch^k \tag{3.28}$$

where

$$
\begin{aligned}
y &= \text{true solution}\\
u &= \text{pp-approximation}\\
C &= \text{a constant}\\
h &= \text{uniform partition}\\
\|Q\| &= \max_x |Q|
\end{aligned}
$$

provided that y is sufficently smooth. Obviously, one can increase the accuracy by choosing the approximating space to be of higher order.

EXAMPLE 2

An insulated metal rod is exposed at each end to a temperature, T_0. Within the rod, heat is generated according to the following function:

$$\xi[(T - T_0) + \cosh(1)]$$

where

$$
\begin{aligned}
\xi &= \text{constant}\\
T &= \text{absolute temperature}
\end{aligned}
$$

The rod is illustrated in Figure 3.4. The temperature profile in the rod can be calculated by solving the following energy balance:

$$
\begin{aligned}
K\frac{d^2T}{dz^2} &= \xi[(T - T_0) + \cosh(1)]\\
T &= T_0 \quad \text{at} \quad z = 0\\
T &= T_0 \quad \text{at} \quad z = L
\end{aligned}
\tag{3.29}
$$

where K is the thermal conductivity of the metal. When $(\xi L^2)/K = 4$, the solution of the BVP is

$$y = \cosh(2x - 1) - \cosh(1)$$

where $y = T - T_0$ and $x = z/L$. Solve (3.29) using the Hermite cubic basis, and show that the order of accuracy is $0(h^4)$ (as expected from 3.28).

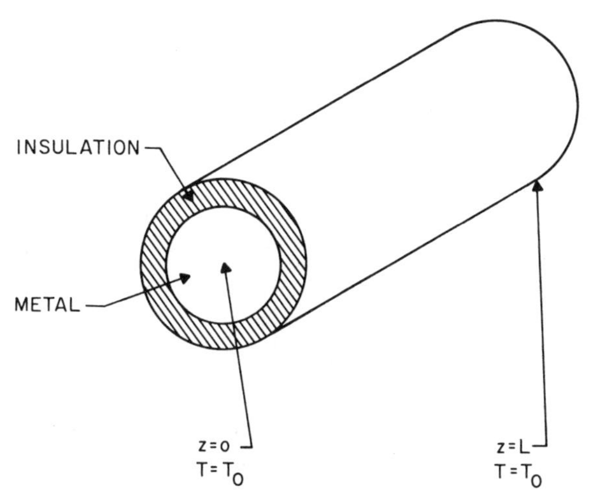

FIGURE 3.4 Insulated metal rod.

SOLUTION

First put (3.29) in dimensionless form by using $y = T - T_0$ and $x = z/L$.

$$\frac{d^2y}{dx^2} = \frac{\xi L^2}{K}[y + \cosh(1)]$$

Since $(\xi L^2)/K = 4$, the ordinary differential equation (ODE) becomes

$$\frac{d^2y}{dx^2} = 4[y + \cosh(1)]$$

Using $\mathscr{P}_4^2(\pi)$ (with π uniform) gives the piecewise polynomial approximation

$$u(x) = \sum_{j=1}^{\ell+1} [\alpha_j^{(1)} v_j + \alpha_j^{(2)} s_j]$$

As with Example 1, $y(0) = y(1) = 0$ and, since $v_1(0) = 1$ and $v_{\ell+1}(1) = 1$,

$$u(x) = \alpha_1^{(2)} s_1 + \alpha_2^{(1)} v_2 + \alpha_2^{(2)} s_2, \ldots, \alpha_\ell^{(1)} v_\ell + \alpha_\ell^{(2)} s_\ell + \alpha_{\ell+1}^{(2)} s_{\ell+1}$$

The weak from of the ODE is

$$-(y', \phi_i') - 4(y, \phi_i) = 4(1, \phi_i)\cosh(1), \qquad i = 1, \ldots, 2(\ell + 1) - 2$$

Substitution of $u(x)$ into the above equation results in

$$-(u', \phi_i') - 4(u, \phi_i) = 4(1, \phi_i)\cosh(1), \qquad i = 1, \ldots, 2(\ell + 1) - 2$$

In matrix notation the previous equation is

$$[\boldsymbol{A} + 4\boldsymbol{B}]\,\boldsymbol{\alpha} = -4\cosh(1)\mathbf{F}$$

where

$$A = \begin{bmatrix} (s_1',s_1') & (s_1',v_2') & (s_1',s_2') & & & & & 0 \\ (v_2',s_1') & (v_2',v_2') & (v_2',s_2') & (v_2',v_3') & (v_2',s_3') & & & \\ (s_2',s_1') & (s_2',v_2') & (s_2',s_2') & (s_2',v_3') & (s_2',s_3') & & & \\ & \cdot & \cdot & \cdot & \cdot & \cdot & & \\ & & \cdot & & \cdot & & \cdot & \\ 0 & & (v_\ell',v_{\ell-1}') & (v_\ell',s_{\ell-1}') & (v_\ell',v_\ell') & (v_\ell',s_\ell') & (v_\ell',s_{\ell+1}') \\ & & (s_\ell',v_{\ell-1}') & (s_\ell',s_{\ell-1}') & (s_\ell',v_\ell') & (s_\ell',s_\ell') & (s_\ell',s_{\ell+1}') \\ & & & & (s_{\ell+1}',v_\ell') & (s_{\ell+1}',s_\ell') & (s_{\ell+1}',s_{\ell+1}') \end{bmatrix}$$

$\boldsymbol{B}$ = the same as $\boldsymbol{A}$ except for no primes on the basis functions

$\mathbf{F} = [(1, s_1), (1, v_2), (1, s_2), \ldots, (1, v_\ell), (1, s_\ell), (1, s_{\ell+1})]^T$

$\boldsymbol{\alpha} = [\alpha_1^{(2)}, \alpha_2^{(1)}, \alpha_2^{(2)}, \ldots, \alpha_\ell^{(1)}, \alpha_\ell^{(2)}, \alpha_{\ell+1}^{(2)}]^T$

Each of the inner products (,) shown in $\boldsymbol{A}$, $\boldsymbol{B}$, and $\mathbf{F}$ must be evaluated. For example

$$v_i(x) = g_1(\xi_{i-1}(x)) + g_1(1 - \xi_i(x))$$

with

$$\xi_{i-1}(x) = \frac{x - x_{i-1}}{x_i - x_{i-1}}, \qquad 1 - \xi_i(x) = \frac{x_{i+1} - x}{x_{i+1} - x_i}$$

Therefore

$$v_i(x) = -2\delta_1 \left[\frac{x - x_{i-1}}{x_i - x_{i-1}}\right]^3 + 3\delta_1 \left[\frac{x - x_{i-1}}{x_i - x_{i-1}}\right]^2 - 2\delta_2 \left[\frac{x_{i+1} - x}{x_{i+1} - x_i}\right]^3 + 3\delta_2 \left[\frac{x_{i+1} - x}{x_{i+1} - x_i}\right]^2$$

where

$$\delta_1 = \begin{cases} 1, & x_{i-1} \leq x \leq x_i \\ 0, & \text{otherwise} \end{cases}$$

$$\delta_2 = \begin{cases} 1, & x_i \leq x \leq x_{i+1} \\ 0, & \text{otherwise} \end{cases}$$

and

$$(1, v_i) = \frac{h_i}{2} + \frac{h_{i+1}}{2}$$

or for a uniform partition,

$$(1, v_i) = h$$

Once all the inner products are determined, the matrix problem is ready to be solved. Notice the structure of $\boldsymbol{A}$ or $\boldsymbol{B}$ (they are the same). These matrices are block-tridiagonal and can be solved using a well-known block version of Gaussian elimination (see page 196 of [3]). The results are shown below.

t	h (uniform partition)	$\|y - u\|$
1	0.1250	0.6011×10^{-5}
2	0.0556	0.2707×10^{-6}
3	0.0357	0.4872×10^{-7}
4	0.0263	0.1475×10^{-7}

Since $\|y - u\| \leqslant Ch^p$, take the logarithm of this equation to give

$$\ln\|y - u\| \leqslant \ln C + pLnh$$

Let $e(h) = \|y - u\|$ (u calculated with a uniform partition; subinterval size h), and calculate p by

$$p = \frac{\ln\left(\dfrac{e(h_{t-1})}{e(h_t)}\right)}{\ln\left(\dfrac{h_{t-1}}{h_t}\right)}$$

From the above results,

t	p
1	—
2	3.83
3	3.87
4	3.91

which shows the fourth-order accuracy of the method.

Thus using $\mathscr{P}_4^2(\pi)$ as the approximating space gives a Galerkin solution possessing a continuous first derivative that is fourth-order accurate.

Nonlinear Equations

Consider the nonlinear ODE:

$$\begin{gathered} y'' = f(x, y, y'), \qquad 0 < x < 1 \\ y(0) = y(1) = 0 \end{gathered} \tag{3.30}$$

Using the B-spline basis gives the pp-approximation

$$u(x) = \sum_{j=1}^{N} \alpha_j B_j(x) \tag{3.31}$$

Substitution of (3.31) into the weak form of (3.30) yields

$$\left(\sum_{j=1}^{N} \alpha_j B_j', B_i'\right) + \left(f\left(x, \sum_{j=1}^{N} \alpha_j B_j, \sum_{j=1}^{N} \alpha_j B_j'\right), B_i\right) = 0, \tag{3.32}$$

$$i = 1, \ldots, N$$

The system (3.32) can be written as

$$A\boldsymbol{\alpha} + \mathbf{H}(\boldsymbol{\alpha}) = \mathbf{0} \tag{3.33}$$

where the vector $\mathbf{H}$ contains inner products that are nonlinear functions of $\boldsymbol{\alpha}$. Equation (3.33) can be solved using Newton's method, but notice that the vector $\mathbf{H}$ must be recomputed after each iteration. Therefore, the computation of $\mathbf{H}$ must be done efficiently. Normally, the integrals in $\mathbf{H}$ do not have closed form, and one must resort to numerical quadrature. The rule of thumb in this case is to use at least an equal number of quadrature points as the degree of the approximating space.

Inhomogeneous Dirichlet and Flux Boundary Conditions

The Galerkin procedures discussed in the previous sections may easily be modified to treat boundary conditions other than the homogeneous Dirichlet conditions, that is, $y(0) = y(1) = 0$. Suppose that the governing ODE is

$$(a(x)y'(x))' + b(x)y(x) + c(x) = 0, \qquad 0 < x < 1 \tag{3.34}$$

subject to the boundary conditions

$$y(0) = \psi_1, \qquad y(1) = \psi_2 \tag{3.35}$$

where ψ_1 and ψ_2 are constants. The weak form of (3.34) is

$$a(x)y'(x)B_i(x)\Big|_0^1 - (a(x)y'(x), B_i'(x)) + (b(x)y(x), B_i(x))$$

$$+ (c(x), B_i(x)) = 0 \tag{3.36}$$

Since

$$B_1(0) = 1, \qquad \sum_{j=2}^{N} B_j(0) = 0$$

and

$$B_N(1) = 1, \qquad \sum_{j=1}^{N-1} B_j(1) = 0$$

then

$$\alpha_1 = \psi_1, \qquad \alpha_N = \psi_2 \tag{3.37}$$

to match the boundary conditions. The value of i in (3.36) goes from 2 to $N - 1$ so that the basis functions satisfy the homogeneous Dirichlet conditions [eliminates the first term in (3.36)]. Thus (3.36) becomes:

$$\sum_{j=2}^{N-1} \alpha_j[(a(x)B_j', B_i') - (b(x)B_j, B_i)] = (c(x), B_i) + \psi_1[-(a(x)B_1', B_i') + (b(x)B_1, B_i)] + \psi_2[-(a(x)B_N', B_i') + (b(x)B_N, B_i)], \quad i = 2, \ldots, N-1 \tag{3.38}$$

If flux conditions are prescribed, they can be represented by

$$\begin{aligned} \eta_1 y + \beta_1 y' &= \gamma_1 \quad \text{at} \quad x = 0 \\ \eta_2 y + \beta_2 y' &= \gamma_2 \quad \text{at} \quad x = 1 \end{aligned} \tag{3.39}$$

where η_1, η_2, β_1, β_2, γ_1, and γ_2 are constants and satisfy

$$|\eta_1| + |\beta_1| > 0$$

$$|\eta_2| + |\beta_2| > 0$$

Write (3.39) as

$$\begin{aligned} y' &= \frac{\gamma_1}{\beta_1} - \frac{\eta_1}{\beta_1} y \quad \text{at} \quad x = 0 \\ y' &= \frac{\gamma_2}{\beta_2} - \frac{\eta_2}{\beta_2} y \quad \text{at} \quad x = 1 \end{aligned} \tag{3.40}$$

Incorporation of (3.40) into (3.36) gives:

$$\sum_{j=1}^{N} \alpha_j \left[(a(x)B_j', B_i') - (b(x)B_j, B_i) - \delta_{i1}\delta_{j1}a(0)\frac{\eta_1}{\beta_1} + \delta_{iN}\delta_{jN}a(1)\frac{\eta_2}{\beta_2} \right] = (c(x), B_i) + \delta_{iN}a(1)\frac{\gamma_2}{\beta_2} - \delta_{i1}a(0)\frac{\gamma_1}{\beta_1}, \quad i = 1, \ldots, N \tag{3.41}$$

where

$$\delta_{st} = \begin{cases} 1, & s = t \\ 0, & s \neq t \end{cases}$$

Notice that the subscript i now goes from 1 to N, since $y(0)$ and $y(1)$ are unknowns.

Mathematical Software

In light of the fact that Galerkin methods are not frequently used to solve BVPs (because of the computational effort as compared with other methods, e.g., finite differences, collocation), it is not surprising that there is very limited software that implements Galerkin methods for BVPs. Galerkin software for BVPs consists of Schryer's code in the PORT library developed at Bell Laboratories [4]. There is a significant amount of Galerkin-based software for partial differential equations, and we will discuss these codes in the chapters concerning partial differential equations. The purpose for covering Galerkin methods for BVPs is for ease of illustration, and because of the straightforward extension into partial differential equations.

COLLOCATION

Consider the nonlinear ODE

$$y'' = f(x, y, y'), \qquad a < x < b \tag{3.42a}$$

$$\eta_1 y + \beta_1 y' = \gamma_1 \quad \text{at} \quad x = a$$

$$\eta_2 y + \beta_2 y' = \gamma_2 \quad \text{at} \quad x = b \tag{3.42b}$$

where η_1, η_2, β_1, β_2, γ_1, and γ_2 are constants. Let the interval partition be given by (3.13), and let the pp-approximation in $\mathscr{P}_k^\nu(\pi)$ ($\nu \geq 2$) be (3.31). The collocation method determines the unknown set $\{\alpha_j|_j = 1, \ldots, N\}$ by satisfying the ODE at N points. For example, if $k = 4$ and $\nu = 2$, then $N = 2\ell + 2$. If we satisfy the two boundary conditions (3.42b), then two collocation points are required in each of the ℓ subintervals. It can be shown that the optimal position of the collocation points are the $k - M$ (M is the degree of the ODE; in this case $M = 2$) Gaussian points given by [5]:

$$\tau_{ji} = x_j + \frac{h_j}{2} + \left(\frac{h_j}{2}\right)\omega_i, \qquad j = 1, \ldots, \ell, \quad i = 1, \ldots, k - M \tag{3.43}$$

where

$$\omega = k - M \text{ Gaussian points in } [-1, 1]$$

The $k - M$ Gaussian points in $[-1, 1]$ are the zeros of the Legendre polynomial of degree $k - M$. For example, if $k = 4$ and $M = 2$, then the two Gaussian points are the zeros of the Legendre polynomial

$$-\tfrac{1}{2} + \tfrac{3}{2}x^2, \qquad -1 \leq x \leq 1$$

or

$$\omega_1 = -\frac{1}{\sqrt{3}}, \qquad \omega_2 = \frac{1}{\sqrt{3}}$$

Thus, the two collocation points in each subinterval are given by

$$\tau_{j1} = x_j + \frac{h_j}{2} - \frac{h_j}{2}\left(\frac{1}{\sqrt{3}}\right)$$

$$\tau_{j2} = x_j + \frac{h_j}{2} + \frac{h_j}{2}\left(\frac{1}{\sqrt{3}}\right) \tag{3.44}$$

The 2ℓ equations specified at the collocation points combined with the two boundary conditions completely determines the collocation solution $\{a_j | j = 1, \ldots, 2\ell + 2\}$.

EXAMPLE 3

Solve Example 2 using spline collocation at the Gaussian points and the Hermite cubic basis. Show the order of accuracy.

SOLUTION

The governing ODE is:

$$\frac{d^2y}{dx^2} = 4[y + \cosh(1)], \qquad 0 < x < 1$$
$$y(0) = y(1) = 0$$

Let

$$Ly = -y'' + 4y = -4 \cosh(1)$$

and consider a general subinterval $[x_j, x_{j+1}]$ in which there are four basis functions—v_j, v_{j+1}, s_j, and s_{j+1}—that are nonzero. The "value" functions are evaluated as follows:

$$v_j = g_1(1 - \xi_j(x)), \qquad [x_j, x_{j+1}]$$

$$v_j = -2\left[\frac{x_{j+1} - x}{h}\right]^3 + 3\left[\frac{x_{j+1} - x}{h}\right]^2, \qquad h = x_{j+1} - x_j$$

$$v_j'' = -\frac{12}{h^3}(x_{j+1} - x) + \frac{6}{h^2}$$

$$Lv_j = \frac{12}{h^3}(x_{j+1} - x) - \frac{6}{h^2} - \frac{8}{h^3}(x_{j+1} - x)^3 + \frac{12}{h^2}(x_{j+1} - x)^2$$

$$v_{j+1} = g_1(\xi_j(x)), \qquad [x_j, x_{j+1}]$$

$$v_{j+1} = -2\left[\frac{x - x_j}{h}\right]^3 + 3\left[\frac{x - x_j}{h}\right]^2$$

$$v_{j+1}'' = -\frac{12}{h^3}(x - x_j) + \frac{6}{h^2}$$

$$Lv_{j+1} = \frac{12}{h^3}(x - x_j) - \frac{6}{h^2} - \frac{8}{h^3}(x - x_j)^3 + \frac{12}{h^2}(x - x_j)^2$$

The two collocation points per subinterval are

$$\tau_{j1} = x_j + \frac{h}{2} - \frac{h}{2}\left(\frac{1}{\sqrt{3}}\right) = x_j + \frac{h}{2}\left[1 - \frac{1}{\sqrt{3}}\right] = x_j + hp_1$$

$$\tau_{j2} = x_j + \frac{h}{2} + \frac{h}{2}\left(\frac{1}{\sqrt{3}}\right) = x_j + \frac{h}{2}\left[1 - \frac{1}{\sqrt{3}}\right] = x_j + hp_2$$

Using τ_{j1} and τ_{j2} in Lv_j and Lv_{j+1} gives

$$
\begin{aligned}
Lv_j(\tau_{j1}) &= \frac{12}{h^2}(1-\rho_1) - \frac{6}{h^2} - 8(1-\rho_1)^3 + 12(1-\rho_1)^2 \\
Lv_j(\tau_{j2}) &= \frac{12}{h^2}(1-\rho_2) - \frac{6}{h^2} - 8(1-\rho_2)^3 + 12(1-\rho_2)^2 \\
Lv_{j+1}(\tau_{j1}) &= \frac{12}{h^2}\rho_1 - \frac{6}{h^2} - 8\rho_1^3 + 12\rho_1^2 \\
Lv_{j+1}(\tau_{j2}) &= \frac{12}{h^2}\rho_2 - \frac{6}{h^2} - 8\rho_2^3 + 12\rho_2^2
\end{aligned}
$$

The same procedure can be used for the "slope" functions to produce

$$
\begin{aligned}
Ls_j(\tau_{j1}) &= \frac{6}{h}(1-\rho_1) - \frac{2}{h} - 4h[(1-\rho_1)^3 - (1-\rho_1)^2] \\
Ls_j(\tau_{j2}) &= \frac{6}{h}(1-\rho_2) - \frac{2}{h} - 4h[(1-\rho_2)^3 - (1-\rho_2)^2] \\
Ls_{j+1}(\tau_{j1}) &= -\frac{6}{h}\rho_1 + \frac{2}{h} + 4h\left[\rho_1^3 - \rho_1^2\right], \\
Ls_{j+1}(\tau_{j2}) &= -\frac{6}{h}\rho_2 + \frac{2}{h} + 4h\left[\rho_2^3 - \rho_2^2\right].
\end{aligned}
$$

For notational convenience let

$$
\begin{aligned}
F1 &= Ls_j(\tau_{j1}) = -Ls_{j+1}(\tau_{j2}) \\
F2 &= Ls_j(\tau_{j2}) = -Ls_{j+1}(\tau_{j1}) \\
F3 &= Lv_j(\tau_{j2}) = Lv_{j+1}(\tau_{j1}) \\
F4 &= Lv_j(\tau_{j1}) = Lv_{j+1}(\tau_{j2})
\end{aligned}
$$

At $x = 0$ and $x = 1$, $y = 0$. Therefore,

$$\alpha_1^{(1)} = \alpha_{\ell+1}^{(1)} = 0$$

Thus the matrix problem becomes:

$$
\begin{bmatrix}
F1 & F3 & -F2 & & & & & & \\
F2 & F4 & -F1 & & & & & & \\
 & F4 & F1 & F3 & -F2 & & 0 & & \\
 & F3 & F2 & F4 & -F1 & & & & \\
 & & \cdot & & & & & & \\
 & & & \cdot & & & & & \\
 & & & & \cdot & & & & \\
 & & & & & F4 & F1 & -F2 \\
 & 0 & & & & F3 & F2 & -F1
\end{bmatrix}
\begin{bmatrix}
\alpha_1^{(2)} \\ \alpha_2^{(1)} \\ \alpha_2^{(2)} \\ \alpha_3^{(1)} \\ \cdot \\ \cdot \\ \cdot \\ \alpha_\ell^{(2)} \\ \alpha_{\ell+1}^{(2)}
\end{bmatrix}
= -4\cosh(1)
\begin{bmatrix}
1 \\ 1 \\ \cdot \\ \cdot \\ \cdot \\ \\ \\ 1 \\ 1
\end{bmatrix}
$$

This matrix problem was solved using the block version of Gaussian elimination (see page 196 of [4]). The results are shown below.

t	h (uniform partition)	$\|y - u\|$
1	0.100	0.2830×10^{-6}
2	0.050	0.1764×10^{-7}
3	0.033	0.3483×10^{-8}
4	0.250	0.1102×10^{-8}

From the above results

t	p
1	—
2	4.00
3	3.90
4	4.14

which shows fourth-order accuracy.

In the previous example we showed that when using $\mathscr{P}_k^\nu(\pi)$, the error was $0(h^4)$. In general, the collocation method using $\mathscr{P}_k^\nu(\pi)$ gives an error of the same order as that in Galerkin's method [Eq. (3.28)] [5].

EXAMPLE 4

The problem of predicting diffusion and reaction in porous catalyst pellets was discussed in Chapter 2. In that discussion the boundary condition at the surface was the specification of a known concentration. Another boundary condition that can arise at the surface of the pellet is the continuity of flux of a species as a result of the inclusion of a boundary layer around the exterior of the pellet. Consider the problem of calculating the concentration profile in an isothermal catalyst pellet that is a slab and is surrounded by a boundary layer. The conservation of mass equation is

$$D \frac{d^2c}{dx^2} = k\mathscr{R}(c), \qquad 0 < x < x_p$$

where

D = diffusivity

x = spatial coordinate (x_p = half thickness of the plate)

c = concentration of a given species

k = rate constant

$\mathscr{R}(c)$ = reaction rate function

The boundary conditions for this equation are

$$\frac{dc}{dx} = 0 \quad \text{at} \quad x = 0 \qquad \text{(symmetry)}$$

$$-D\frac{dc}{dx} = S_h(c - c_0) \quad \text{at} \quad x = x_p \qquad \text{(continuity of flux)}$$

where

c_0 = known concentration at the exterior of the boundary layer

S_h = mass transfer coefficient

Set up the matrix problem to solve this ODE using collocation with $\mathcal{P}_4^2(\overline{\pi})$, where

$$\overline{\pi}: \quad 0 = x_1 < x_2 < \ldots < x_{\ell+1} = x_p$$

and

$$h = x_{i+1} - x_i, \qquad \text{for } 1 \leqslant i \leqslant \ell \quad \text{(i.e., uniform)}$$

SOLUTION

First, put the conservation of mass equation in dimensionless form by defining

$$C = \frac{c}{c_0}$$

$$z = \frac{x}{x_p}$$

$$\Phi = x_p\sqrt{\frac{k}{D}} \qquad \text{(Thiele modulus)}$$

$$\text{Bi} = \frac{S_h x_p}{D} \qquad \text{(Biot number)}$$

With these definitions, the ODE becomes

$$\frac{d^2C}{dz^2} = \Phi^2\left[\frac{\mathcal{R}(c)}{c_0}\right]$$

$$\frac{dC}{dz} = 0 \quad \text{at} \quad z = 0$$

$$\frac{dC}{dz} = \text{Bi}\,(1 - C) \quad \text{at} \quad z = 1$$

The dimension of $\mathscr{P}_4^2(\overline{\pi})$ is $2(\ell + 1)$, and there are two collocation points in each subinterval.

The pp-approximation is

$$u(x) = \sum_{j=1}^{\ell+1} (\alpha_j^{(1)} v_j + \alpha_j^{(2)} s_j)$$

With $C'(0) = 0$, $\alpha_1^{(2)}$ is zero since $s_1' = 1$ is the only nonzero basis function in $u'(0)$. For each subinterval there are two equations such that

$$\sum_{j=1}^{\ell+1} \left[\alpha_j^{(1)} v_j''(\tau_{i1}) + \alpha_j^{(2)} s_j''(\tau_{i1})\right] = \frac{\Phi^2}{c_0} \mathscr{R} \left\{ c_0 \left[\sum_{j=1}^{\ell+1} \alpha_j^{(1)} v_j(\tau_{i1}) + \alpha_j^{(2)} s_j(\tau_{i1}) \right] \right\}$$

$$\sum_{j=1}^{\ell+1} \left[\alpha_j^{(1)} v_j''(\tau_{i2}) + \alpha_j^{(2)} s_j''(t_{i2})\right] = \frac{\Phi^2}{c_0} \mathscr{R} \left\{ c_0 \left[\sum_{j=1}^{\ell+1} \alpha_j^{(1)} v_j(\tau_{i2}) + \alpha_j^{(2)} s_j(\tau_{i2}) \right] \right\}$$

for $i = 1, \ldots, \ell$. At the boundary $z = 1$ we have

$$\alpha_{\ell+1}^{(2)} = \text{Bi}\,(1 - \alpha_{\ell+1}^{(1)})$$

since $s_{\ell+1}' = 1$ is the only nonzero basis function in $u'(1)$ and $v_{\ell+1} = 1$ is the only nonzero basis funciton in $u(1)$.

Because the basis is local, the equations at the collocation points can be simplified. In matrix notation:

$$\begin{bmatrix} v_1''(\tau_{11}),\ v_2''(\tau_{11}),\ s_2''(\tau_{11}) & & & \\ v_1''(\tau_{12}),\ v_2''(\tau_{12}),\ s_2''(\tau_{12}) & & & \\ & v_2''(\tau_{21}),\ s_2''(\tau_{21}),\ v_3''(\tau_{21}),\ s_3''(\tau_{21}) & & 0 \\ & v_2''(\tau_{22}),\ s_2''(\tau_{22}),\ v_3''(\tau_{22}),\ s_3''(\tau_{22}) & & \\ & \cdot & & \\ & & \cdot & \\ & & \cdot & \\ & & \cdot & \\ 0 & & v_\ell''(\tau_{\ell 1}),\ s_\ell''(\tau_{\ell 1}),\ v_{\ell+1}''(\tau_{\ell 1}) - \text{Bi}\ s_{\ell+1}''(\tau_{\ell 1}) & \\ & & v_\ell''(\tau_{\ell 2}),\ s_\ell''(\tau_{\ell 2}),\ v_{\ell+1}''(\tau_{\ell 2}) - \text{Bi}\ s_{\ell+1}''(\tau_{\ell 2}) & \end{bmatrix} \begin{bmatrix} \alpha_1^{(1)} \\ \alpha_2^{(1)} \\ \alpha_2^{(2)} \\ \alpha_3^{(1)} \\ \alpha_3^{(2)} \\ \cdot \\ \cdot \\ \cdot \\ \alpha_\ell^{(1)} \\ \alpha_\ell^{(2)} \\ \alpha_{\ell+1}^{(1)} \end{bmatrix} = \frac{\Phi^2}{c_0} \mathbf{F}$$

where

$$\mathbf{F} = \begin{bmatrix} \mathscr{R}\{c_0[\alpha_1^{(1)}v_1(\tau_{11}) + \alpha_2^{(1)}v_2(\tau_{11}) + \alpha_2^{(2)}s_2(\tau_{11})]\} \\ \vdots \\ \mathscr{R}\{c_0[\alpha_j^{(1)}v_j(\tau_{j1}) + \alpha_j^{(2)}s_j(\tau_{j1}) + \alpha_{j+1}^{(1)}v_{j+1}(\tau_{j1}) + \alpha_{j+1}^{(2)}s_{j+1}(\tau_{j1})]\} \\ \mathscr{R}\{c_0[\alpha_j^{(1)}v_j(\tau_{j2}) + \alpha_j^{(2)}s_j(\tau_{j2}) + \alpha_{j+1}^{(1)}v_{j+1}(\tau_{j2}) + \alpha_{j+1}^{(2)}s_{j+1}(\tau_{j2})]\} \\ \vdots \\ -\dfrac{c_0\mathrm{Bi}}{\Phi^2}s''_{\ell+1}(\tau_{\ell 1}) + \mathscr{R}\{c_0[\alpha_\ell^{(1)}v_\ell(\tau_{\ell 1}) + \alpha_\ell^{(2)}s_\ell(\tau_{\ell 1}) \\ \qquad + \alpha_{\ell+1}^{(1)}(v_{\ell+1}(\tau_{\ell 1}) - \mathrm{Bi}s_{\ell+1}(\tau_{\ell 1})) + \mathrm{Bi}s_{\ell+1}(\tau_{\ell 1})]\} \\ -\dfrac{c_0\mathrm{Bi}}{\Phi^2}s''_{\ell+1}(\tau_{\ell 2}) + \mathscr{R}\{c_0[\alpha_\ell^{(1)}v_\ell(\tau_{\ell 2}) + \alpha_\ell^{(2)}s_\ell(\tau_{\ell 2}) \\ \qquad + \alpha_{\ell+1}^{(1)}(v_{\ell+1}(\tau_{\ell 2}) - \mathrm{Bi}s_{\ell+1}(\tau_{\ell 2})) + \mathrm{Bi}s_{\ell+1}(\tau_{\ell 2})]\} \end{bmatrix}$$

This problem is nonlinear, and therefore Newton's method or a variant of it would be used. At each iteration the linear system of equations can be solved efficiently by the alternate row and column elimination procedure of Varah [6]. This procedure has been modified and a FORTRAN package was produced by Diaz et al. [7].

As a final illustration of collocation, consider the m nonlinear ODEs

$$\mathbf{y}'' = \mathbf{f}(x, \mathbf{y}, \mathbf{y}'), \qquad a < x < b \tag{3.45a}$$

with

$$\mathbf{g}(\mathbf{y}(a), \mathbf{y}(b), \mathbf{y}'(a), \mathbf{y}'(b)) = \mathbf{0} \tag{3.45b}$$

The pp-approximations ($\mathscr{P}_k^\nu(\pi)$) for this system can be written as

$$\mathbf{u}(x) = \sum_{j=1}^{N} \boldsymbol{\alpha}_j B_j(x) \tag{3.46}$$

where each $\boldsymbol{\alpha}_j$ is a constant vector of length m. The collocation equations for (3.45) are

$$\sum_{j=1}^{N} \boldsymbol{\alpha}_j B''_j(\tau_{si}) = \mathbf{f}\left(x, \sum_{j=1}^{N} \boldsymbol{\alpha}_j B_j(\tau_{si}), \sum_{j=1}^{N} \boldsymbol{\alpha}_j B'_j(\tau_{si})\right)$$
$$i = 1, \ldots, k-2, \qquad s = 1, \ldots, \ell \tag{3.47}$$

and,

$$\mathbf{g}\left(\sum_{j=1}^{N} \boldsymbol{\alpha}_j B_j(a), \sum_{j=1}^{N} \boldsymbol{\alpha}_j B_j(b), \sum_{j=1}^{N} \boldsymbol{\alpha}_j B'_j(a), \sum_{j=1}^{N} \boldsymbol{\alpha}_j B'_j(b)\right) = \mathbf{0} \tag{3.48}$$

If there are m ODEs in the system and the dimension of $\mathscr{P}_k^\nu(\pi)$ is N, then there are mN unknown coefficients that must be obtained from the nonlinear algebraic system of equations composed of (3.47) and (3.48). From (3.23)

$$N = k + \sum_{j=2}^{\ell} (k - \nu) \tag{3.49}$$

and the number of coefficients is thus

$$mk + m(\ell - 1)(k - \nu) \tag{3.50}$$

The number of equations in (3.47) is $m(k - 2)\ell$, and in (3.48) is $2m$. Therefore the system (3.47) and (3.48) is composed of $2m + m\ell(k - 2)$ equations. If we impose continuity of the first derivative, that is, $\nu = 2$, then (3.50) becomes

$$mk + m(\ell - 1)(k - 2)$$

or

$$2m + m\ell(k - 2) \tag{3.51}$$

Thus the solution of the system (3.47) and (3.48) completely specifies the pp-approximation.

Mathematical Software

The available software that is based on collocation is rather limited. In fact, it consists of one code, namely COLSYS [8]. Next, we will study this code in detail.

COLSYS uses spline collocation to determine the solution of the mixed-order system of equations

$$u_s^{(M_s)}(x) = f_s(x; \mathbf{z}(\mathbf{u})), \qquad s = 1, \ldots, d$$

$$a < x < b \tag{3.52}$$

where

$$\begin{aligned} M_s &= \text{order of the } s \text{ differential equation} \\ \mathbf{u} &= [u_1, u_2, \ldots, u_d]^T \text{ is the vector of solutions} \\ \mathbf{z}(\mathbf{u}) &= (u_1, u_1', \ldots, u_1^{(M_1-1)}, \ldots, u_d, u_d', \ldots, u_d^{(M_d-1)}) \end{aligned}$$

It is assumed that the components $u_1, u_2, \ldots, u_d$ are ordered such that

$$M_1 \leqslant M_2 \leqslant \ldots \leqslant M_d \leqslant 4 \tag{3.53}$$

Equations (3.52) are solved with the conditions

$$g_j(\xi_j, \mathbf{z}(\mathbf{u})) = 0, \qquad j = 1, \ldots, M^* \tag{3.54}$$

where

$$M^* = \sum_{s=1}^{d} M_s$$

and

$$a \leqslant \xi_i \leqslant \xi_2 \leqslant \ldots \leqslant \xi_{M^*} \leqslant b$$

Unlike the BVP codes in Chapter 2, COLSYS does not convert (3.52) to a first-order system. While (3.54) does not allow for nonseparated boundary conditions, such problems can be converted to the form (3.54) [9]. For example, consider the BVP

$$y'' = f(x, y, y'), \qquad a < x < b$$
$$y'(a) = \alpha, \qquad g(y(a), y(b)) = 0 \tag{3.55}$$

Introducing a (constant) $V(x)$ gives an equivalent BVP

$$y'' = f(x, y, y'), \qquad a < x < b$$
$$V' = 0, \tag{3.56}$$
$$y'(a) = \alpha, \qquad y(a) = V(a), \qquad g(V(b), y(b)) = 0$$

which does not contain a nonseparated boundary condition.

COLSYS implements the method of spline collocation at Gaussian points using a B-spline basis (modified versions of deBoor's algorithms [2] are used to calculate the B-splines and their derivates). The pp-solutions are thus in $\mathscr{P}_k^{\nu^*}(\pi)$ where COLSYS sets k and ν^* such that

$$u_s \text{ is in } \mathscr{P}_{q+M_s}^{\nu^*}(\pi), \qquad s = 1, \ldots, d \tag{3.57}$$

where

$$\nu^* = \{\nu_j = M_s \mid j = 2, \ldots, \ell\}$$

$$\text{q} = \text{number of collocation points per subintervals}$$

The matrix problem is solved using an efficient implementation of Gaussian elimination with partial pivoting [10], and nonlinear problems are "handled" by the use of a modified Newton method. Algorithms are included for estimating the error, and for mesh refinement. A redistribution of mesh points is automatically performed (if deemed worthwhile) to roughly equidistribute the error. This code has proven to be quite effective for the solution of "difficult" BVPs arising in chemical engineering [11].

To illustrate the use of COLSYS we will solve the isothermal effectiveness factor problem with large Thiele moduli. The governing BVP is the conservation of mass in a porous plate catalyst pellet where a second-order reaction rate is occurring, i.e.,

$$\frac{d^2c}{dx^2} = \Phi^2 c^2, \qquad 0 < x < 1,$$
$$c'(0) = 0$$
$$c(1) = 1 \tag{3.58}$$

where

c = dimensionless concentration of a given species

x = dimensionless coordinate

Φ = Thiele modulus (constant)

The effectiveness factor (defined in Chapter 2) for this problem is

$$E = \int_0^1 c^2 \, dx \tag{3.59}$$

For large values of Φ, the "exact" solution can be obtained [12] and is

$$E = \frac{1}{\Phi} \sqrt{\frac{2}{3}} \, (1 - c_0^3)^{1/2} \tag{3.60}$$

where c_0 is given by

$$\Phi \sqrt{\frac{2}{3} c_0} = \int_0^{1/c_0} \frac{d\xi}{\sqrt{\xi^3 - 1}} \tag{3.61}$$

This problem is said to be difficult because of the extreme gradient in the solution (see Figure 3.5). We now present the results generated by COLSYS.

COLSYS was used to solve (3.58) with $\Phi = 50$, 100, and 150. A tolerance was set on the solution and the first derivative, and an initial uniform mesh of five subintervals was used with initial solution and derivative profiles of 0.1 and 0.001 for $0 \leqslant x \leqslant 1$, respectively. The solution for $\Phi = 50$ was used as the initial profile for calculating the solution with $\Phi = 100$, and subsequently this solution was used to calculate the solution for $\Phi = 150$. Table 3.3 compares

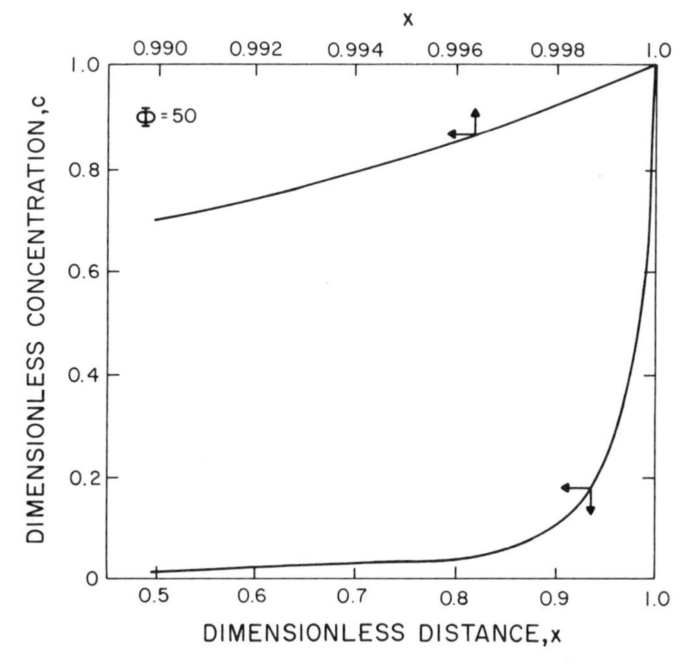

FIGURE 3.5 Solution of Eq. (3.58).

TABLE 3.3 Results for Eq. (3.58)
Tolerance = 10^{-4}
Collocation Points Per Subinterval = 3

Φ	COLSYS	"Exact"
50	0.1633(−1)	0.1633(−1)
100	0.8165(−2)	0.8165(−2)
150	0.5443(−2)	0.5443(−2)

the results computed by COLSYS with those of (3.60) and (3.61). This table shows that COLSYS is capable of obtaining accurate results for this "difficult" problem.

COLSYS incorporates an error estimation and mesh refinement algorithm. Figure 3.6 shows the redistribution of the mesh for $\Phi = 50$, $q = 4$, and the tolerance = 10^{-4}. With the initial uniform mesh (mesh redistribution number = 0; i.e., a mesh redistribution number of η designates that COLSYS has automatically redistributed the mesh η times), COLSYS performed eight Newton iterations on the matrix problem to achieve convergence. Since the computations continued, the error exceeded the specified tolerance. Notice that the mesh was then redistributed such that more points are placed in the region of the steep gradient (see Figure 3.5). This is done to "equidistribute" the error throughout the x interval. Three additional redistributions of the mesh were required to provide an approximation that met the specified error tolerance. Finally, the effect of the tolerance and q, the number of collocation points per subinterval, were tested. In Table 3.4, one can see the results of varying the aforementioned parameters. In all cases shown, the same solution, $u(x)$, and value of E were

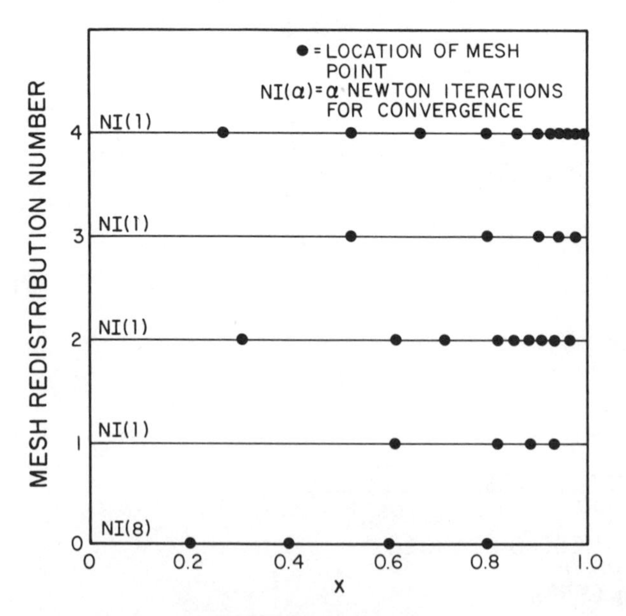

FIGURE 3.6 Redistribution of mesh.

TABLE 3.4 Further Results for Eq. (3.58)
$\Phi = 50$

Collocation Points Per Subinterval	Number of Subintervals	Tolerance on Solution and Derivative	E.T.R.*
3	20	10^{-4}	1.0
3	114	10^{-6}	4.6
2	80	10^{-4}	1.9
4	12	10^{-4}	1.1

* E.T.R. = execution time ratio.

obtained. As the tolerance is lowered, the number of subintervals and the execution time required for solution increase. This is not unexpected since we are asking the code to calculate a more accurate solution. When q is raised from 3 to 4, there is a slight decrease in the number of subintervals required for solution, but this requires approximately the same execution time. If q is reduced from 3 to 2, notice the quadrupling in the number of subintervals used for solution and also the approximate doubling of the execution time. The drastic changes in going from $q = 2$ to $q = 3$ and the relatively small changes when increasing q from 3 to 4 indicate that for this problem one should specify $q \geqslant 3$.

In this chapter we have outlined two finite element methods and have discussed the limited software that implements these methods. The extension of these methods from BVPs to partial differential equations is shown in later chapters.

PROBLEMS

1. A liquid is flowing in laminar motion down a vertical wall. For $z < 0$, the wall does not dissolve in the fluid, but for $0 < z < L$, the wall contains a species A that is slightly soluble in the liquid (see Figure 3.7, from [13]). In this situation, the change in the mass convection in the z direction equals the change in the diffusion of mass in the x direction, or

$$\frac{\partial}{\partial z}(u_z c_A) = D\frac{\partial^2 c_A}{\partial x^2}$$

where u_z is the velocity and D is the diffusivity. For a short "contact time" the partial differential equation becomes (see page 561 of [13]):

$$\hat{a}x\frac{\partial c_A}{\partial z} = D\frac{\partial^2 c_A}{\partial x^2}$$

$$c_A = 0 \quad \text{at} \quad z = 0$$

$$c_A = 0 \quad \text{at} \quad x = \infty$$

$$c_A = c_A^0 \quad \text{at} \quad x = 0$$

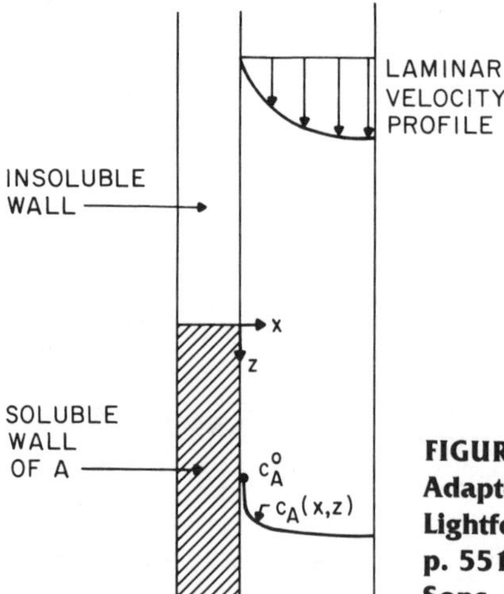

FIGURE 3.7 Solid dissolution into falling film. Adapted from R. B. Bird, W. E. Stewart, and E. N. Lightfoot, *Transport Phenomena,* copyright © 1960, p. 551. Reprinted by permission of John Wiley and Sons, New York.

where $\hat{a}$ is a constant and c_A^0 is the solubility of A in the liquid. Let

$$f = \frac{c_A}{c_A^0} \quad \text{and} \quad \xi = x\left(\frac{\hat{a}}{9Dz}\right)^{1/3}$$

The PDE can be transformed into a BVP with the use of the above dimensionless variables:

$$\frac{d^2f}{d\xi^2} + 3\xi^2\frac{df}{d\xi} = 0$$

$$f = 0 \quad \text{at} \quad \xi = \infty$$

$$f = 1 \quad \text{at} \quad \xi = 0$$

Solve this BVP using the Hermite cubic basis by Galerkin's method and compare your results with the closed-form solution (see p. 552 of [13]):

$$f = \frac{\int_\xi^\infty \exp(-\xi^3)d\xi}{\Gamma(\frac{4}{3})}$$

where $\Gamma(n) = \int_0^\infty \beta^{n-1}e^{-\beta}d\beta$, $(n > 0)$, which has the recursion formula

$$\Gamma(n + 1) = n\Gamma(n)$$

The solution of the Galerkin matrix problem should be performed by calling an appropriate matrix routine in a library available at your installation.

2. Solve Problem 1 using spline collocation at Gaussian points.

3.* Solve problem 5 of Chapter 2 and compare your results with those obtained with a discrete variable method.

4.* The following problem arises in the study of compressible boundary layer flow at a point of attachment on a general curved surface [14].

$$f''' + (f + cg)f'' + (1 + S_w h - (f')^2) = 0$$

$$g''' + (f + cg)g'' + c(1 + S_w h - (f')^2) = 0$$

$$h'' + (f + cg)h' = 0$$

with

$$f = g = f' = g' = 0 \quad \text{at} \quad \eta = 0$$

$$h = 1 \quad \text{at} \quad \eta = 0$$

$$f' = g' = 1 \quad \text{at} \quad \eta \to \infty$$

$$h = 0 \quad \text{at} \quad \eta \to \infty$$

where f, g, and h are functions of the independent variable η, and c and S_w are constants. As initial approximations use

$$f(\eta) = g(\eta) = \frac{\eta^2}{2\eta_\infty}$$

$$h(\eta) = \frac{\eta_\infty - \eta}{\eta_\infty}$$

where η_∞ is the point at which the right-hand boundary conditions are imposed. Solve this problem with $S_w = 0$ and $c = 1.0$ and compare your results with those given in [11].

5.* Solve Problem 4 with $S_w = 0$ and $c = -0.5$. In this case there are two solutions. Be sure to calculate both solutions.

6.* Solve Problem 3 with $\beta = 0$ but allow for a boundary layer around the exterior of the pellet. The boundary condition at $x = 1$ now becomes

$$\frac{dy}{dx} = \text{Bi}\,(1 - y)$$

Vary the value of Bi and explain the effects of the boundary layer.

REFERENCES

1. Fairweather, G., *Finite Element Galerkin Methods for Differential Equations,* Marcel Dekker, New York (1978).

2. deBoor, C., *Practical Guide to Splines,* Springer-Verlag, New York (1978).

3. Varga, R. S., *Matrix Iterative Analysis,* Prentice-Hall, Englewood Cliffs, N.J. (1962).
4. Fox, P. A., A. D. Hall, and N. L. Schryer, "The PORT Mathematical Subroutine Library," ACM TOMS, *4,* 104 (1978).
5. deBoor, C., and B. Swartz, "Collocation at Gaussian Points," SIAM J. Numer. Anal., *10,* 582 (1973).
6. Varah, J. M., "Alternate Row and Column Elimination for Solving Certain Linear Systems," SIAM J. Numer. Anal., *13,* 71 (1976).
7. Diaz, J. C., G. Fairweather, and P. Keast, "FORTRAN Packages for Solving Almost Block Diagonal Linear Systems by Alternate Row and Column Elimination," Tech. Rep. No. 148/81, Department of Computer Science, Univ. Toronto (1981).
8. Ascher, U., J. Christiansen, and R. D. Russell, "Collocation Software for Boundary Value ODEs," ACM TOMS, *7,* 209 (1981).
9. Ascher, U., and R. D. Russell, "Reformulation of Boundary Value Problems Into "Standard" Form," SIAM Rev. *23,* 238 (1981).
10. deBoor, C., and R. Weiss, "SOLVEBLOK: A Package for Solving Almost Block Diagonal Linear Systems," ACM TOMS, *6,* 80 (1980).
11. Davis, M., and G. Fairweather, "On the Use of Spline Collocation for Boundary Value Problems Arising in Chemical Engineering," Comput. Methods. Appl. Mech. Eng., *28,* 179 (1981).
12. Aris, R., *The Mathematical Theory of Diffusion and Reaction in Permeable Catalysts,* Clarendon Press, Oxford (1975).
13. Bird, R. B., W. E. Stewart, and E. N. Lightfoot, *Transport Phenomena,* Wiley, New York (1960).
14. Poots, J., "Compressible Laminar Boundary-Layer Flow at a Point of Attachment," J. Fluid Mech., *22,* 197 (1965).

BIBLIOGRAPHY

For additional or more detailed information, see the following:

Becker, E. B., G. F. Carey, and J. T. Oden, *Finite Elements: An Introduction,* Prentice-Hall, Englewood Cliffs, N.J. (1981).

deBoor, C., *Practical Guide to Splines,* Springer-Verlag, New York (1978).

Fairweather, G., *Finite Element Galerkin Methods for Differential Equations,* Marcel Dekker, New York (1978).

Russell, R. D., *Numerical Solution of Boundary Value Problems,* Lecture Notes, Universidad Central de Venezuela, Publication 79-06, Caracas (1979).

Strang, G., and G. J. Fix, *An Analysis of the Finite Element Method,* Prentice-Hall, Englewood Cliffs, N.J. (1973).

4

Parabolic Partial Differential Equations in One Space Variable

INTRODUCTION

In Chapter 1 we discussed methods for solving IVPs, whereas in Chapters 2 and 3 boundary-value problems were treated. This chapter combines the techniques from these chapters to solve parabolic partial differential equations in one space variable.

CLASSIFICATION OF PARTIAL DIFFERENTIAL EQUATIONS

Consider the most general linear partial differential equation of the second order in two independent variables:

$$Lw = aw_{xx} + 2bw_{xy} + cw_{yy} + dw_x + ew_y + fw = g \tag{4.1}$$

where a, b, c, d, e, f, g are given functions of the independent variables and the subscripts denote partial derivatives. The principal part of the operator L is

$$a \frac{\partial^2}{\partial x^2} + 2b \frac{\partial^2}{\partial x \, \partial y} + c \frac{\partial^2}{\partial y^2} \tag{4.2}$$

Primarily, it is the principal part, (4.2), that determines the properties of the solution of the equation $Lw = g$. The partial differential equation $Lw - g = 0$

is classified as:

$$\left.\begin{array}{l}\text{Hyperbolic}\\\text{Parabolic}\\\text{Elliptic}\end{array}\right\}\quad\text{according as}\quad b^2 - ac\quad\left\{\begin{array}{l}> 0\\= 0\\< 0\end{array}\right. \tag{4.3}$$

where $b^2 - ac$ is called the discriminant of L. The procedure for solving each type of partial differential equation is different. Examples of each type are:

$$w_{xx} + w_{yy} = 0, \quad \text{e.g., Laplace's equation, which is elliptic}$$

$$w_{xx} - c^2 w_{yy} = 0, \quad \text{e.g., wave equation, which is hyperbolic}$$

$$w_x - D w_{yy} = 0, \quad \text{e.g., diffusion equation, which is parabolic}$$

An equation can be of mixed type depending upon the values of the parameters, e.g.,

$$\begin{array}{c} y w_{xx} + w_{yy} = 0,\\ \text{(Tricomi's equation)}\end{array} \quad \left\{\begin{array}{l} y < 0, \text{ hyperbolic}\\ y = 0, \text{ parabolic,}\\ u > 0, \text{ elliptic}\end{array}\right.$$

To each of the equations (4.1) we must adjoin appropriate subsidiary relations, called boundary and/or initial conditions, which serve to complete the formulation of a "meaningful problem." These conditions are related to the domain in which (4.1) is to be solved.

METHOD OF LINES

Consider the diffusion equation:

$$\frac{\partial w}{\partial t} = D\frac{\partial^2 w}{\partial x^2}, \qquad 0 < x < 1, \quad 0 < t$$

$$D = \text{constant}, \tag{4.4}$$

with the following mesh in the x-direction

$$h = x_{i+1} - x_i, \qquad i = 1, \ldots, N$$

$$x_1 = 0, \qquad x_{N+1} = 1 \tag{4.5}$$

Discretize the spatial derivative in (4.4) using finite differences to obtain the following system of ordinary differential equations:

$$\frac{du_i}{dt} = \frac{D}{h^2}\left[u_{i+1} - 2u_i + u_{i-1}\right] \tag{4.6}$$

where

$$u_i(t) \simeq w(x_i, t)$$

Thus, the parabolic PDE can be approximated by a coupled system of ODEs in t. This technique is called the method of lines (MOL) [1] for obvious reasons. To complete the formulation we require knowledge of the subsidiary conditions. The parabolic PDE (4.4) requires boundary conditions at $x = 0$ and $x = 1$, and an initial condition at $t = 0$. Three types of boundary conditions are:

$$\begin{aligned} &\text{Dirichlet,} \quad && \text{e.g., } w(0, t) = g_1(t) \\ &\text{Neumann,} \quad && \text{e.g., } w_x(1, t) = g_2(t) \\ &\text{Robin,} \quad && \text{e.g., } \alpha w(0, t) + \beta w_x(0, t) = g_3(t) \end{aligned} \tag{4.7}$$

In the MOL, the boundary conditions are incorporated into the discretization in the x-direction while the initial condition is used to start the associated IVP.

EXAMPLE 1

Write down the MOL discretization for

$$\frac{\partial w}{\partial t} = D\frac{\partial^2 w}{\partial x^2}$$

$$w(0, t) = \alpha$$

$$w(1, t) = \beta$$

$$w(x, 0) = \alpha + (\beta - \alpha)x$$

using a uniform mesh, where D, α, and β are constants.

SOLUTION

Referring to (4.6), we have

$$\frac{du_i}{dt} = \frac{D}{h^2}[u_{i+1} - 2u_i + u_{i-1}], \qquad i = 2, \ldots, N$$

For $i = 1$, $u_1 = \alpha$, and for $i = N + 1$, $u_{N+1} = \beta$ from the boundary conditions. The ODE system is therefore:

$$\frac{du_2}{dt} = \frac{1}{h^2}[u_3 - 2u_2 + \alpha]$$

$$\frac{du_i}{dt} = \frac{1}{h^2}[u_{i+1} - 2u_i + u_{i-1}], \qquad i = 3, \ldots, N - 1$$

$$\frac{du_N}{dt} = \frac{1}{h^2}[\beta - 2u_N + u_{N-1}]$$

with

$$u_i = \alpha + (\beta - \alpha)x_i \quad \text{at} \quad t = 0$$

This IVP can be solved using the techniques discussed in Chapter 1.

The method of lines is a very useful technique since IVP solvers are in a more advanced stage of development than other types of differential equation solvers. We have outlined the MOL using a finite difference discretization. Other discretization alternatives are finite element methods such as collocation and Galerkin methods. In the following sections we will first examine the MOL using finite difference methods, and then discuss finite element methods.

FINITE DIFFERENCES

Low-Order Time Approximations

Consider the diffusion equation (4.4) with

$$\begin{aligned} w(0, t) &= 0 \\ w(1, t) &= 0 \\ w(x, 0) &= f(x) \end{aligned}$$

which can represent the unsteady-state diffusion of momentum, heat, or mass through a homogeneous medium. Discretize (4.4) using a uniform mesh to give:

$$\begin{aligned} \frac{du_i}{dt} &= \frac{D}{h^2}[u_{i+1} - 2u_i + u_{i-1}], \qquad i = 2, \ldots, N \\ u_1 &= 0 \\ u_{N+1} &= 0 \end{aligned} \tag{4.8}$$

where $u_i = f(x_i)$, $i = 2, \ldots, N$ at $t = 0$. If the Euler method is used to integrate (4.8), then with

$$\begin{aligned} u_{i,j} &\simeq w(x_i, t_j) \\ t_j &= j\,\Delta t, \qquad j = 0, 1, \ldots \end{aligned}$$

we obtain

$$\frac{u_{i,j+1} - u_{i,j}}{\Delta t} = \frac{D}{h^2}[u_{i+1,j} - 2_{i,j} + u_{i-1,j}] \tag{4.9}$$

or

$$u_{i,j+1} = (1 - 2\tau)u_{i,j} + \tau(u_{i+1,j} + u_{i-1,j})$$

where

$$\tau = \frac{\Delta t}{h^2} D$$

and the error in this formula is $0(\Delta t + h^2)$ (Δt from the time discretization, h^2 from the spatial discretization). At $j = 0$ all the u_i's are known from the initial condition. Therefore, implementation of (4.9) is straightforward:

1. Calculate $u_{i,j+1}$ for $i = 2, \ldots, N$ (u_1 and u_{N+1} are known from the boundary conditions), using (4.9) with $j = 0$.
2. Repeat step (1) using the computed $u_{i,j+1}$ values to calculate $u_{i,j+2}$, and so on.

Equation (4.9) is called the forward difference method.

EXAMPLE 2

Calculate the solution of (4.8) with

$$D = 1$$

$$f(x) = \begin{cases} 2x, & \text{for } 0 \leq x \leq \frac{1}{2} \\ 2(1 - x), & \text{for } \frac{1}{2} \leq x \leq 1 \end{cases}$$

Use $h = 0.1$ and let (1) $\Delta t = 0.001$, and (2) $\Delta t = 0.01$.

SOLUTION

Equation (4.9) with $h = 0.1$ and $\Delta t = 0.001$ becomes:

$$u_{i,j+1} = 0.8u_{i,j} + 0.1(u_{i+1,j} + u_{i-1,j}) \qquad (\tau = 0.1)$$

The solution of this equation at $x = 0.1$ and $t = 0.001$ is

$$u_{2,1} = 0.8u_{2,0} + 0.1(u_{3,0} + u_{1,0})$$

The initial condition gives

$$u_{2,0} = 2h$$

$$u_{3,0} = 2(2h)$$

$$u_{1,0} = 0$$

Thus, $u_{2,1} = 0.16 + 0.04 = 0.2$. Likewise $u_{3,1} = 0.4$. Using $u_{3,1}$, $u_{2,1}$, $u_{1,1}$, one can then calculate $u_{2,2}$ as

$$u_{2,2} = 0.2.$$

A sampling of some results are listed below:

t	**Finite-Difference Solution ($x = 0.3$)**	**Analytical Solution ($x = 0.3$)**
0.005	0.5971	0.5966
0.01	0.5822	0.5799
0.02	0.5373	0.5334
0.10	0.2472	0.2444

Now solve (4.9) using $h = 0.1$ and $\Delta t = 0.01$ ($\tau = 1$). The results are:

	x					
t	**0.0**	**0.1**	**0.2**	**0.3**	**0.4**	**0.5**
0.00	0	0.2	0.4	0.6	0.8	1.0
0.01	0	0.2	0.4	0.6	0.8	0.6
0.02	0	0.2	0.4	0.6	0.4	1.0
0.03	0	0.2	0.4	0.2	1.2	−0.2
0.04	0	0.2	0.0	1.4	−1.2	2.6

As one can see, the computed results are very much affected by the choice of τ.

In Chapter 1 we saw that the Euler method had a restrictive stability criterion. The analogous behavior is shown in the forward difference method. The stability criterion for the forward difference method is [2]:

$$\tau \leqslant \tfrac{1}{2} \tag{4.10}$$

As with IVPs there are two properties of PDEs that motivate the derivation of various algorithms, namely stability and accuracy. Next we discuss a method that has improved stability properties.

Consider again the discretization of (4.4), i.e., (4.8). If the implicit Euler method is used to integrate (4.8), then (4.8) is approximated by

$$\frac{u_{i,j+1} - u_{i,j}}{\Delta t} = \frac{D}{h^2}\left[u_{i+1,j+1} - 2u_{i,j+1} + u_{i-1,j+1}\right] \tag{4.11}$$

or

$$u_{i,j} = -\tau u_{i+1,j+1} + (1 + 2\tau)u_{i,j+1} - \tau u_{i-1,j+1}$$

The error in (4.11) is again $0(\Delta t + h^2)$. Notice that in contrast to (4.9), (4.11) is implicit. Therefore, denote

$$\mathbf{u}_{j+1} = (u_{2,j+1}, u_{3,j+1}, \ldots, u_{N,j+1})^T \tag{4.12}$$

and write (4.11) in matrix notation as:

$$\begin{bmatrix} 1+2\tau & -\tau & & & \\ -\tau & 1+2\tau & -\tau & & \\ & \cdot & \cdot & \cdot & \\ & & \cdot & \cdot & \cdot \\ & & \cdot & \cdot & \cdot \\ & & & \cdot & -\tau \\ & & & -\tau & 1+2\tau \end{bmatrix} \mathbf{u}_{j+1} = \mathbf{u}_j + \tau \begin{bmatrix} u_{1,j+1} \\ 0 \\ \cdot \\ \cdot \\ \cdot \\ 0 \\ u_{N+1,j+1} \end{bmatrix} \tag{4.13}$$

with $u_{1,j+1} = u_{N+1,i+1} = 0$. Equation (4.11) is called the backward difference method, and it is unconditionally stable. One has gained stability over the forward difference method at the expense of having to solve a tridiagonal linear system, but the same accuracy is maintained.

To achieve higher accuracy in time, discretize the time derivative using the trapezoidal rule:

$$\left.\frac{\partial w}{\partial t}\right|_{j+1/2} = \frac{w_{i,j+1} - w_{i,j}}{\Delta t} + 0(\Delta t^2) \tag{4.14}$$

where

$$w_{i,j} = w(x_i, t_j)$$

Notice that (4.14) requires the differential equation to be approximated at the half time level. Therefore the spatial discretization must be at the half time level. If the average of $w_{i,j}$ and $w_{i,j+1}$ is used to approximate $w_{i,j+1/2}$, then (4.4) becomes

$$\frac{w_{i,j+1} - w_{i,j}}{\Delta t} = \frac{D}{2h^2}[(w_{i+1,j+1} + w_{i+1,j}) - 2(w_{i,j+1} + w_{i,j})$$
$$+ (w_{i-1,j+1} + w_{i-1,j})] + 0(h^2 + \Delta t^2) \tag{4.15}$$

A numerical procedure for the solution of (4.4) can be obtained from (4.15) by truncating $0(\Delta t^2 + h^2)$ and is:

$$\begin{bmatrix} 1+\tau & -\tau/2 & & & \\ -\tau/2 & 1+\tau & -\tau/2 & & \\ \cdot & \cdot & \cdot & & \\ & \cdot & \cdot & \cdot & \\ & & \cdot & \cdot & \cdot \\ & & & \cdot & \cdot & -\tau/2 \\ & & & & -\tau/2 & 1+\tau \end{bmatrix} \mathbf{u}_{j+1} = \begin{bmatrix} 1-\tau & \tau/2 & & & \\ \tau/2 & 1-\tau & \tau/2 & & \\ \cdot & \cdot & \cdot & & \\ & \cdot & \cdot & \cdot & \\ & & \cdot & \cdot & \cdot \\ & & & \cdot & \cdot & \tau/2 \\ & & & & \tau/2 & 1-\tau \end{bmatrix} \mathbf{u}_j - \begin{bmatrix} \frac{\tau}{2}(u_{1,j} + u_{1,j+1}) \\ \cdot \\ 0 \\ \cdot \\ \cdot \\ \frac{\tau}{2}(u_{N+1,j} + u_{N+1,j+1}) \end{bmatrix} \tag{4.16}$$

where $u_{1,j}$, $u_{1,j+1}$, $u_{N+1,j}$ and $u_{N+1,j+1} = 0$. This procedure is called the Crank-Nicolson method, and it is unconditionally stable [2].

The Theta Method

The forward, backward, and Crank-Nicolson methods are special cases of the theta method. In the theta method the spatial derivatives are approximated by the following combination at the j and $j + 1$ time levels:

$$\frac{\partial^2 u}{\partial x^2} = \theta \delta_x^2 u_{i,j+1} + (1 - \theta)\delta_x^2 u_{i,j} \tag{4.17}$$

where

$$\delta_x^2 u_{i,j} = \frac{u_{i+1,j} - 2u_{i,j} + u_{i-1,j}}{h^2}$$

For example, (4.4) is approximated by

$$\frac{u_{i,j+1} - u_{i,j}}{\Delta t} = D[\theta \delta_x^2 u_{i,j+1} + (1 - \theta)\delta_x^2 u_{i,j}] \tag{4.18}$$

or in matrix form

$$[\boldsymbol{I} + \theta \tau \boldsymbol{J}]\, \mathbf{u}_{j+1} = [\boldsymbol{I} - (1 - \theta)\tau \boldsymbol{J}]\mathbf{u}_j \tag{4.19}$$

where

$$\boldsymbol{I} = \text{identity matrix} \tag{4.20}$$

$$\boldsymbol{J} = \begin{bmatrix} 2 & -1 & & & \\ -1 & 2 & \cdot & & \\ & \cdot & \cdot & \cdot & \\ & & \cdot & \cdot & -1 \\ & & & -1 & 2 \end{bmatrix}$$

Referring to (4.18), we see that $\theta = 0$ is the forward difference method and the spatial derivative is evaluated at the jth time level. The computational molecule for this method is shown in Figure 4.1a. For $\theta = 1$ the spatial derivative is evaluated at the $j + 1$ time level and its computational molecule is shown in

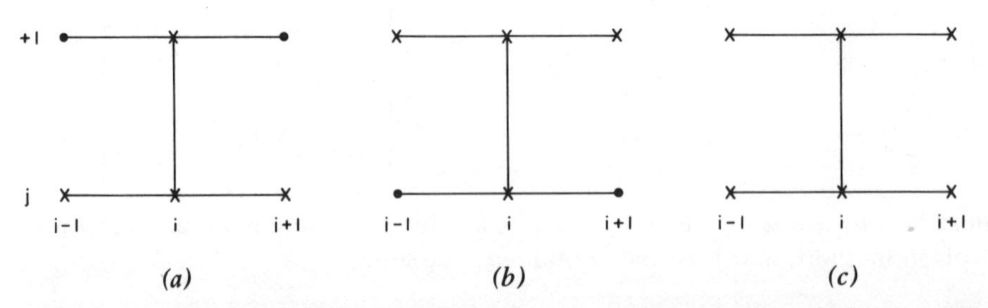

FIGURE 4.1 Computation molecules (x denoted grid points involved in the difference formulation). (a) Forward-difference method. (b) Backward-difference method. (c) Crank-Nicolson method.

Figure 4.1*b*. The Crank-Nicolson method approximates the differential equation at the $j + \frac{1}{2}$ time level (computational molecule shown in Figure 4.1*c*) and requires information from six positions. Since $\theta = \frac{1}{2}$, the Crank-Nicolson method averages the spatial derivative between j and $j + 1$ time levels. Theta may lie anywhere between zero and one, but for values other than 0, 1, and $\frac{1}{2}$ there is no direct correspondence with previously discussed methods. Equation (4.19) can be written conveniently as:

$$\mathbf{u}_{j+1} = [\boldsymbol{I} + \theta\tau\boldsymbol{J}]^{-1}[\boldsymbol{I} - (1 - \theta)\tau\boldsymbol{J}]\,\mathbf{u}_j \tag{4.21}$$

or

$$\mathbf{u}_{j+1} = \boldsymbol{c}\mathbf{u}_j = [\boldsymbol{c}]^j\mathbf{u}_1 \tag{4.22}$$

Boundary and Initial Conditions

Thus far, we have only discussed Dirichlet boundary conditions. Boundary conditions expressed in terms of derivatives (Neumann or Robin conditions) occur very frequently in practice. If a particular problem contains flux boundary conditions, then they can be treated using either of the two methods outlined in Chapter 2, i.e., the method of false boundaries or the integral method. As an example, consider the problem of heat conduction in an insulated rod with heat being convected "in" at $x = 0$ and convected "out" at $x = 1$. The problem can be written as

$$\rho C_p \frac{\partial T}{\partial t} = k\frac{\partial^2 T}{\partial x^2}$$

$$T = T_0 \quad \text{at} \quad t = 0, \qquad \text{for } 0 < x < 1$$

$$-k\frac{\partial T}{\partial x} = h_1(T_1 - T) \quad \text{at} \quad x = 0$$

$$-k\frac{\partial T}{\partial x} = h_2(T - T_2) \quad \text{at} \quad x = 1 \tag{4.23}$$

where

T = dimensionless temperature

T_0 = dimensionless initial temperature

ρC_p = density times the heat capacity of the rod

h_1, h_2 = convective heat transfer coefficients

T_1, T_2 = dimensionless reference temperatures

k = thermal conductivity of the rod

Using the method of false boundaries, at $x = 0$

$$-k \frac{\partial T}{\partial x} = h_1(T_1 - T)$$

becomes

$$-k \frac{u_2 - u_0}{2\,\Delta x} = h_1(T_1 - u_1) \tag{4.24}$$

Solving for u_0 gives

$$u_0 = \frac{2h_1}{k} \Delta x\,(T_1 - u_1) + u_2 \tag{4.25}$$

A similar procedure can be used at $x = 1$ in order to obtain

$$u_{N+2} = \frac{2h_2}{k} \Delta x(T_2 - u_{N+1}) + u_N \tag{4.26}$$

Thus the Crank-Nicolson method for (4.23) can be written as:

$$\mathbf{u}_{j+1} = \mathbf{c}\mathbf{u}_j + \mathbf{f} \tag{4.27}$$

where

$$\mathbf{c} = [I + \tfrac{1}{2}\tau A]^{-1}[\mathbf{I} - \tfrac{1}{2}\tau A]$$

$$A = \begin{bmatrix} 2\left(1 + \Delta x \frac{h_1}{k}\right) & -2 & & & \\ -1 & 2 & -1 & & \\ & \cdot & \cdot & \cdot & \\ & & \cdot & \cdot & \cdot \\ & & \cdot & \cdot & \cdot \\ & & -1 & 2 & -1 \\ & & & -2 & 2\left(1 + \Delta x \frac{h_2}{k}\right) \end{bmatrix}$$

$$\mathbf{u}_j = [u_{1,j}, u_{2,j}, \ldots, u_{N+1,j}]^T$$

$$\mathbf{f} = \left[2\tau\,\Delta x \frac{h_1}{k} T_1, 0, \ldots, 0, 2\tau\,\Delta x \frac{h_2}{k} T_2\right]^T$$

$$\tau = \frac{k\,\Delta t}{\rho C_p(\Delta x)^2}$$

An interesting problem concerning the boundary and initial conditions that can arise in practical problems is the incompatibility of the conditions at some point. To illustrate this effect, consider the problem of mass transfer of a component into a fluid flowing through a pipe with a soluble wall. The situation is

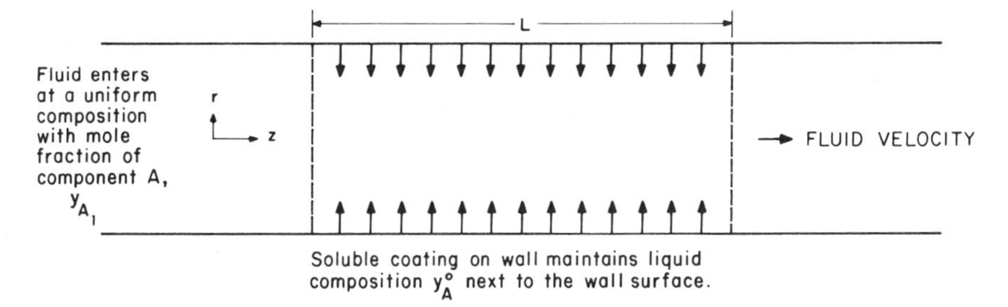

FIGURE 4.2 Mass transfer in a pipe with a soluble wall. Adapted from R. B. Bird, W. E. Stewart, and E. N. Lightfoot, *Transport Phenomena,* copyright © 1960, p. 643. Reprinted by permission of John Wiley and Sons, New York.

shown in Figure 4.2. The governing differential equation is simply a material balance on the fluid

$$\underset{\text{(a)}}{v \frac{\partial y_A}{\partial z}} = \underset{\text{(b)}}{\frac{\mathscr{D}}{r} \frac{\partial}{\partial r}\left(r \frac{\partial y_A}{\partial r}\right)} \tag{4.28a}$$

with

$$y_A = y_{A_1} \quad \text{at} \quad z = 0, \qquad \text{for } 0 \leqslant r \leqslant \text{wall} \tag{4.28b}$$

$$\frac{\partial y_A}{\partial r} = 0 \quad \text{at} \quad r = 0 \tag{4.28c}$$

$$-\mathscr{D} \frac{\partial y_A}{\partial r} = k_g(y_A - y_A^0) \qquad \text{at} \qquad r = \text{wall} \tag{4.28d}$$

where

$$\mathscr{D} = \text{diffusion coefficient}$$

$$v = \text{fluid velocity}$$

$$k_g = \text{mass transfer coefficient}$$

Term (a) is the convection in the z-direction and term (b) is the diffusion in the r-direction. Notice that condition (4.28b) does not satisfy condition (4.28d) at r = wall. This is what is known as inconsistency in the initial and boundary conditions.

The question of the assignment of y_A at $z = 0$, r = wall now arises, and the analyst must make an arbitrary choice. Whatever choice is made, it introduces errors that, if the difference scheme is stable, will decay at successive z levels (a property of stable parabolic equation solvers). The recommendation of Wilkes [3] is to use the boundary condition value and set $y_A = y_A^0$ at $z = 0$, r = wall.

EXAMPLE 3

A fluid (constant density ρ and viscosity μ) is contained in a long horizontal pipe of length L and radius R. Initially, the fluid is at rest. At $t = 0$, a pressure gradient $(P_0 - P_L)/L$ is imposed on the system. Determine the unsteady-state velocity profile V as a function of time.

SOLUTION

The governing differential equation is

$$\rho \frac{\partial V}{\partial t} = \frac{P_0 - P_L}{L} + \mu \frac{1}{r} \frac{\partial}{\partial r}\left(r \frac{\partial V}{\partial r}\right)$$

with

$$V = 0 \quad \text{at} \quad t = 0, \qquad \text{for } 0 \leqslant r \leqslant R$$

$$\frac{\partial V}{\partial r} = 0 \quad \text{at} \quad r = 0, \qquad \text{for } t \geqslant 0$$

$$V = 0 \quad \text{at} \quad r = R, \qquad \text{for } t \geqslant 0$$

Define

$$\tau = \frac{\mu t}{\rho R^2}$$

$$\xi = \frac{r}{R}$$

$$\eta = \frac{4\mu L V}{(P_0 - P_L)R^2}$$

then the governing PDE can be written as

$$\frac{\partial \eta}{\partial \tau} = 4 + \frac{1}{\xi} \frac{\partial}{\partial \xi}\left(\xi \frac{\partial \eta}{\partial \xi}\right)$$

$$\eta = 0 \quad \text{at} \quad \tau = 0, \qquad \text{for } 0 \leqslant \xi \leqslant 1$$

$$\frac{\partial \eta}{\partial \xi} = 0 \quad \text{at} \quad \xi = 0, \text{ for } \tau \geqslant 0$$

$$\eta = 0 \quad \text{at} \quad \xi = 1, \text{ for } \tau \geqslant 0$$

At $\tau \rightarrow \infty$ the system attains steady state, η_∞. Let

$$\eta(\tau, \xi) = \eta_\infty(\xi) - \phi(\tau, \xi)$$

The steady-state solution is obtained by solving

$$0 = 4 + \frac{1}{\xi} \frac{d}{d\xi}\left(\xi \frac{d\eta_\infty}{d\xi}\right)$$

with

$$\eta_\infty = 0 \quad \text{at} \quad \xi = 1$$

$$\frac{\partial \eta_\infty}{\partial \xi} = 0 \quad \text{at} \quad \xi = 0$$

and is $\eta_\infty = 1 - \xi^2$. Therefore

$$\frac{\partial \phi}{\partial \tau} = \frac{1}{\xi}\frac{\partial}{\partial \xi}\left(\xi \frac{\partial \phi}{\partial \xi}\right)$$

with

$$\frac{\partial \phi}{\partial \xi} = 0 \quad \text{at} \quad \xi = 0, \qquad \text{for } \tau \geqslant 0$$

$$\phi = 0 \quad \text{at} \quad \xi = 1, \qquad \text{for } \tau \geqslant 0$$

$$\phi = 1 - \xi^2 \quad \text{at} \quad \tau = 0, \qquad \text{for } 0 \leqslant \xi \leqslant 1$$

Discretizing the above PDE using the theta method yields:

$$[\boldsymbol{I} + \theta\Phi\boldsymbol{A}]\,\mathbf{u}_{j+1} = [\boldsymbol{I} - (1 - \theta)\Phi\boldsymbol{A}]\,\mathbf{u}_j$$

where

$$\Phi = \frac{\Delta\tau}{(\Delta\xi)^2}$$

$$\boldsymbol{A} = \begin{bmatrix} 4 & -4 & & & \\ -\left(1 - \frac{1}{2(i-1)}\right) & 2 & -\left(1 + \frac{1}{2(i-1)}\right) & & \\ & \cdot & \cdot & \cdot & \\ & & \cdot & \cdot & \cdot \\ & & & \cdot & \cdot & \cdot \\ & & & & \cdot & \cdot & \cdot \\ & & & & & -\left(1 - \frac{1}{2(i-1)}\right) & 2 \end{bmatrix}$$

$$u_{i,j} \simeq \phi(\xi_i, \tau_j)$$

Table 4.1 shows the results. For $\Phi > 0.5$, the solution with $\theta = 0$ diverges. As Φ goes from 0.4 to 0.04, the solutions with $\theta = 0, 1$ approach the solution with $\theta = \frac{1}{2}$. Also notice that no change in the $\theta = \frac{1}{2}$ solution occurs when decreasing Φ. Since $\theta = 0, 1$ are $0(\Delta\tau + \Delta\xi^2)$ and $\theta = \frac{1}{2}$ is $0(\Delta\tau^2 + \Delta\xi^2)$, one would expect the behavior shown in Table 4.1. The analytical solution is given for $\tau = 0.2$ and 0.4, but as stated in the table, it required interpolation from Bessel function tables, and therefore is not applicable for a five-digit precision comparison with the numerical results. The analytical solution is given to show that all the answers

TABLE 4.1 Computed ϕ Values Using the Theta Method $\xi = 0.4$

	$\Phi = 0.4$			$\Phi = 0.04$			
τ	$\theta = 0$	$\theta = 1$	$\theta = \frac{1}{2}$	$\theta = 0$	$\theta = 1$	$\theta = \frac{1}{2}$	Analytical†
0.2	0.27192	0.27374	0.27283	0.27274	0.27292	0.27283	0.2723
0.4	0.85394(−1)	0.86541(−1)	0.85967(−1)	0.85910(−1)	0.86025(−1)	0.85968(−1)	0.8567(−1)
0.8	0.84197(−2)	0.86473(−2)	0.85332(−2)	0.85218(−2)	0.85446(−2)	0.85332(−2)	—

† Solution required the use of Bessel functions, and interpolation of tabular data will produce errors in these numbers.

are within "engineering accuracy." Finally, the unsteady-state velocity profile is shown in Figure 4.3, and is what one would expect from the physical situation.

Nonlinear Equations

Consider the nonlinear equation

$$\begin{aligned} w_{xx} &= F(x, t, w, w_x, w_t) \\ 0 \leq x \leq 1, &\qquad 0 \leq t \end{aligned} \tag{4.29}$$

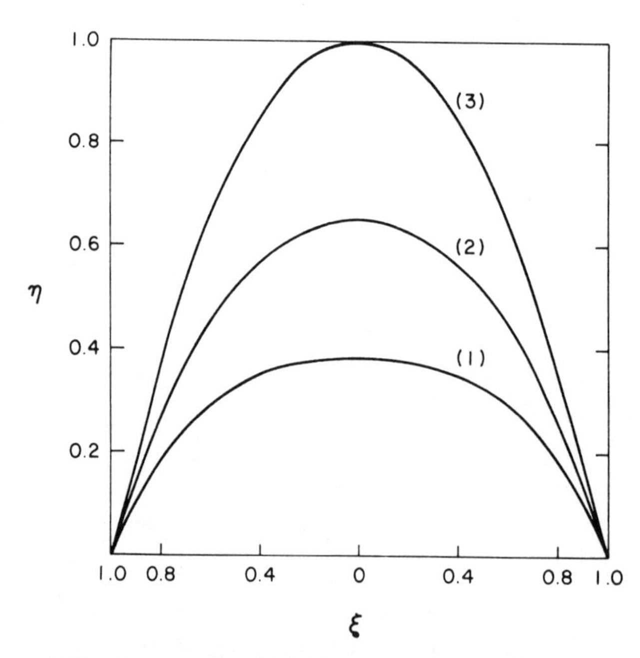

FIGURE 4.3 Transient velocity profiles.

	τ
(1)	0.1
(2)	0.2
(3)	∞

with $w(0, t)$, $w(1, t)$, and $w(x, 0)$ specified. The forward difference method would produce the following difference scheme for (4.29):

$$\delta_x^2 u_{i,j} = F\left(x_i, t_j, u_{i,j}, \Delta_x u_{i,j}, \frac{u_{i,j+1} - u_{i,j}}{\Delta t}\right) \tag{4.30}$$

where

$$\Delta_x u_{i,j} = \frac{u_{i+1,j} - u_{i-1,j}}{2\,\Delta x}$$

If the time derivative appears linearly, (4.30) can be solved directly. This is because all the nonlinearities are evaluated at the jth level, for which the node values are known. The stability criterion for the forward difference method is not the same as was derived for the linear case, and no generalized explicit criterion is available. For "difficult" problems implicit methods should be used. The backward difference and Crank-Nicolson methods are

$$\delta_x^2 u_{i,j+1} = F\left(x_i, t_j, u_{i,j+1}, \Delta_x u_{i,j+1}, \frac{u_{i,j+1} - u_{i,j}}{\Delta t}\right) \tag{4.31}$$

and

$$\frac{1}{2}\,\delta_x^2(u_{i,j+1} + u_{i,j}) = F\left(x_i, t_{j+1/2}, \frac{u_{i,j+1} + u_{i,j}}{2}, \tfrac{1}{2}\Delta_x\,(u_{i,j+1} + u_{i,j}), \frac{u_{i,j+1} - u_{i,j}}{\Delta t}\right) \tag{4.32}$$

Equations (4.31) and (4.32) lead to systems of nonlinear equations that must be solved at each time step. This can be done by a Newton iteration.

To reduce the computation time, it would be advantageous to use methods that handle nonlinear equations without iteration. Consider a special case of (4.29), namely,

$$w_{xx} = f_1(x, t, w)w_t + f_2(x, t, w)w_x + f_3(x, t, w) \tag{4.33}$$

A Crank-Nicolson discretization of (4.33) gives

$$\frac{1}{2}\delta_x^2\,(u_{i,j+1} + u_{i,j}) = f_1^{i,j+1/2}\left(\frac{u_{i,j+1} - u_{i,j}}{\Delta t}\right) + f_2^{i,j+1/2}\,\Delta_x\left(\frac{u_{i,j+1} + u_{i,j}}{2}\right) + f_3^{i,j+1/2} \tag{4.34}$$

with

$$f_n^{i,j+1/2} = f_n\left(x_i, t_{j+1/2}, \frac{u_{i,j+1} + u_{i,j}}{2}\right), \qquad n = 1, 2, 3.$$

Equation (4.34) still leads to a nonlinear system of equations that would require an iterative method for solution. If one could estimate $u_{i,j+1/2}$, by $\tilde{u}_{i,j+1/2}$ say, and use it in

$$\frac{1}{2}\,\delta_x^2\,(u_{i,j+1} + u_{i,j}) = \bar{f}_1^{i,j+1/2}\left(\frac{u_{i,j+1} - u_{i,j}}{\Delta t}\right) + \bar{f}_2^{i,j+1/2}\,\Delta_x\left(\frac{u_{i,j+1} + u_{i,j}}{2}\right) + \bar{f}_3^{i,j+1/2}$$

where

$$\bar{f}_m = f_m(x_i,\ t_{j+1/2},\ \tilde{u}_{i,j+1/2}) \tag{4.35}$$

then there would be no iteration. Douglas and Jones [4] have considered this problem and developed a predictor-corrector method. They used a backward difference method to calculate $\tilde{u}_{i,j+1/2}$:

$$\delta_x^2\,\tilde{u}_{i,j+1/2} = \hat{f}_1^{i,j}\,\frac{\tilde{u}_{i,j+1/2} - u_{i,j}}{\left(\dfrac{\Delta t}{2}\right)} + \hat{f}_2^{i,j}\,\Delta_x\,\tilde{u}_{i,j+1/2} + \hat{f}_3^{i,j},$$

where

$$\hat{f}_m^{i,j} = f_m(x_i,\ t_{j+1/2},\ u_{i,j}) \tag{4.36}$$

The procedure is to predict $\tilde{u}_{i,j+1/2}$ from (4.36) (this requires the solution of one tridiagonal linear system) and to correct using (4.35) (which also requires the solution of one tridiagonal linear system). This method eliminates the nonlinear iteration at each time level, but does require that two tridiagonal systems be solved at each time level. Lees [5] introduced an extrapolated Crank-Nicolson method to eliminate the problem of solving two tridiagonal systems at each time level. A linear extrapolation to obtain $\tilde{u}_{i,j+1/2}$ gives

$$\tilde{u}_{i,j+1/2} = u_{i,j} + \tfrac{1}{2}\,(u_{i,j} - u_{i,j-1})$$

or

$$\tilde{u}_{i,j+1/2} = \frac{3u_{i,j} - u_{i,j-1}}{2} \tag{4.37}$$

Notice that $\tilde{u}_{i,j+1/2}$ is defined directly for $j > 1$. Therefore, the procedure is to calculate the first time level using either a forward difference approximation or the predictor-corrector method of Douglas and Jones, then step in time using (4.35) with $\tilde{u}_{i,j+1/2}$ defined by (4.37). This method requires the solution of only one tridiagonal system at each time level (except for the first step), and thus would require less computation time.

Inhomogeneous Media

Problems containing inhomogeneities occur frequently in practical situations. Typical examples of these are an insulated pipe—i.e., interfaces at the inside fluid–inside pipewall, outside pipewall–inside insulation surface, and the outside insulation surface–outside fluid—or a nuclear fuel element (refer back to Chapter 2). In this case, the derivation of the PDE difference scheme is an extension of that which was outlined in Chapter 2 for ODEs.

Consider the equation

$$\frac{\partial w}{\partial z} = \frac{\partial}{\partial r}\left[A(r)\,\frac{\partial w}{\partial r}\right] \tag{4.38}$$

at the interface shown in Figure 4.4. Let

$$h_{\rm I} = r_i - r_{i-1}$$

$$h_{\rm II} = r_{i+1} - r_i$$

and

$$A(r) = \begin{cases} A_{\rm I}(r), & \text{for } r < r_i \\ A_{\rm II}(r) & \text{for } r > r_i \end{cases} \tag{4.39}$$

with $A(r)$ being discontinuous at r_i. For w continuous at r_i and

$$A_{\rm II}(r_i^+)\,\frac{\partial w}{\partial r}\bigg|_{r_i^+} = A_{\rm I}(r_i^-)\,\frac{\partial w}{\partial r}\bigg|_{r_i^-} \tag{4.40}$$

the discretization of (4.38) at r_i can be formulated as follows. Integrate (4.38) from $r_{i+1/2}$ to r_i:

$$A_{\rm II}(r_{i+1/2})\,\frac{\partial w}{\partial r}\bigg|_{r_{i+1/2}} - A_{\rm II}(r_i^+)\,\frac{\partial w}{\partial R}\bigg|_{r_i^+} = \int_{r_i^+}^{r_{i+1/2}} \frac{\partial w}{\partial z}\,dr \tag{4.41}$$

Next, integrate (4.38) from r_i to $r_{i-1/2}$.

$$A_{\rm I}(r_i^-)\,\frac{\partial w}{\partial r}\bigg|_{r_i^-} - A_{\rm I}(r_{i-1/2})\,\frac{\partial w}{\partial r}\bigg|_{r_{i-1/2}} = \int_{r_{i-1/2}}^{r_i^-} \frac{\partial w}{\partial z}\,dr \tag{4.42}$$

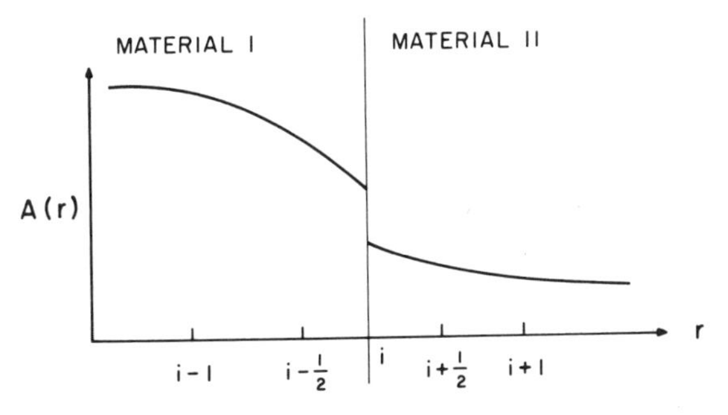

FIGURE 4.4 Material interface where the function $A(r)$ is discontinuous.

Now, add (4.42) to (4.41) and use the continuity condition (4.40) to give

$$A_{\text{II}}(r_{i+1/2})\left.\frac{\partial w}{\partial r}\right|_{r_{i+1/2}} - A_{\text{I}}(r_{i-1/2})\left.\frac{\partial w}{\partial r}\right|_{r_{i-1/2}} = \int_{r_{i-1/2}}^{r_{i+1/2}} \frac{\partial w}{\partial z}\,dr \tag{4.43}$$

Approximate the integral in (4.43) by

$$\int_{r_{i-1/2}}^{r_{i+1/2}} \frac{\partial w}{\partial z}\,dr \simeq \frac{\partial w}{\partial z}(r_{i+1/2} - r_{i-1/2}) = \frac{1}{2}(h_{\text{I}} + h_{\text{II}})\frac{\partial w}{\partial z} \tag{4.44}$$

If a Crank-Nicolson formulation is desired, then (4.43) and (4.44) would give

$$\begin{aligned}\frac{u_{i,j+1} - u_{i,j}}{\Delta z} = \frac{1}{h_{\text{I}} + h_{\text{II}}}\Bigg\{&\frac{A_{\text{II}}(r_{i+1/2})}{h_{\text{II}}}(u_{i+1,j+1} + u_{i+1,j})\\ &- \left(\frac{A_{\text{II}}(r_{i+1/2})}{h_{\text{II}}} + \frac{A_{\text{I}}(r_{i-1/2})}{h_{\text{I}}}\right)(u_{i,j+1} + u_{i,j})\\ &+ \frac{A_{\text{I}}(r_{i-1/2})}{h_{\text{I}}}(u_{i-1,j+1} + u_{i-1,j})\Bigg\}\end{aligned} \tag{4.45}$$

Notice that if $h_{\text{I}} = h_{\text{II}}$ and $A_{\text{I}} = A_{\text{II}}$, then (4.45) reduces to the standard second-order correct Crank-Nicolson discretization. Thus the discontinuity of $A(r)$ is taken into account.

As an example of a problem containing interfaces, we outline the solution of the material balance equation of the annular bed reactor [6]. Figure 4.5 is a schematic of the annular bed reactor, ABR. This reactor is made up of an annular catalyst bed of very small particles next to the heat transfer surface with the inner core of the annulus packed with large, inert spheres (the two beds being separated by an inert screen). The main fluid flow is in the axial direction through

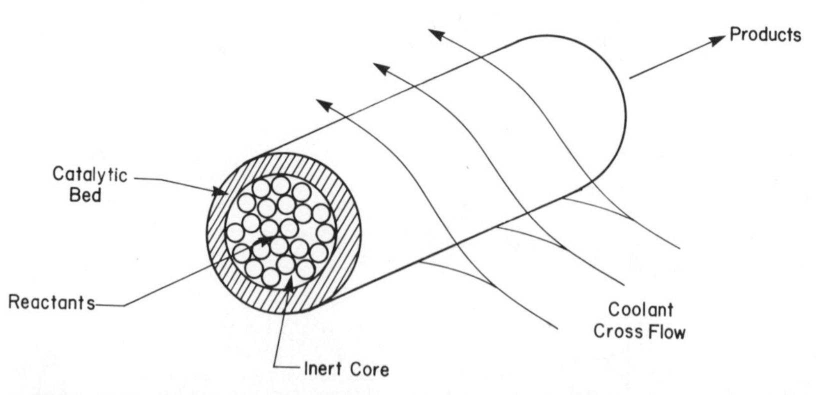

FIGURE 4.5 Schematic of annular bed reactor. From M. E. Davis and J. Yamanis, A.I.Ch.E. J., *28*, p. 267 (1982). Reprinted by permission of the A.I.Ch.E. Journal and the authors.

the core, where the inert packing promotes radial transport to the catalyst bed. If the effects of temperature, pressure, and volume change due to reaction on the concentration and on the average velocity are neglected and mean properties are used, the mass balance is given by

$$\underset{\text{(a)}}{\delta_1 \frac{\partial f}{\partial z}} = \underset{\text{(b)}}{\left[\frac{\text{Am An}}{\text{Re Sc}}\right] \frac{1}{r} \frac{\partial}{\partial r}\left(rD \frac{\partial f}{\partial r}\right)} + \underset{\text{(c)}}{\left[\frac{\text{Am An}}{\text{Re Sc}}\right] \delta_2\, \phi^2 R(f)} \tag{4.46}$$

where the value of 1 or 0 for δ_1 and δ_2 signifies the presence or absence of a term from the above equation as shown below

	Core	**Screen**	**Bed**
δ_1	1	0	0
δ_2	0	0	−1

with

f = dimensionless concentration
z = dimensionless axial coordinate
r = dimensionless radial coordinate
Am, An = aspect ratios (constants)
Re = Reynolds number
Sc = Schmidt number
D = dimensionless radial dispersion coefficient
ϕ = Thiele modulus
$R(f)$ = dimensionless reaction rate function.

Equation (4.46) must be complemented by the appropriate boundary and initial conditions, which are given by

$$\frac{\partial f}{\partial r} = 0 \quad \text{at} \quad r = 0 \qquad \text{(centerline)}$$

$$D^c \left.\frac{\partial f}{\partial r}\right|_c = D^{sc} \left.\frac{\partial f}{\partial r}\right|_{sc} \quad \text{at} \quad r = r_{sc} \qquad \text{(core-screen interface)}$$

$$D^{sc} \left.\frac{\partial f}{\partial r}\right|_{sc} = D^B \left.\frac{\partial f}{\partial r}\right|_B \quad \text{at} \quad r = r_B \qquad \text{(screen-bed interface)}$$

$$\frac{\partial f}{\partial r} = 0 \quad \text{at} \quad r = 1 \qquad \text{(wall)}$$

$$f = 1 \quad \text{at} \quad z = 0, \qquad \text{for } 0 \leq r \leq 1$$

Notice that in (4.46), term (a) represents the axial convection and therefore is not in the equation for the screen and bed regions, while term (c) represents reaction that only occurs in the bed, i.e., the equation changes from parabolic to elliptic when moving from the core to the screen and bed. Also, notice that D is discontinuous at r_{sc} and r_B. This problem is readily solved by the use of an equation of the form (4.45). Equation (4.46) becomes

$$\delta_1\left[\frac{u_{i,j+1} - u_{i,j-1}}{\Delta z}\right] = \left[\frac{\text{Am An}}{\text{Re Sc}}\right]\frac{1}{h_{\text{I}} + h_{\text{II}}}\frac{1}{r_i}\left\{\frac{r_{i+1/2}\,D_{i+1/2}}{h_{\text{II}}}(u_{i+1,j+1}\right.$$

$$+ u_{i+1,j}) - \left(\frac{r_{i+1/2}\,D_{i+1/2}}{h_{\text{II}}} + \frac{r_{i-1/2}\,D_{i-1/2}}{h_{\text{I}}}\right)(u_{i,j+1} + u_{i,j})$$

$$\left. + \frac{r_{i-1/2}\,D_{i-1/2}}{h_{\text{I}}}(u_{i-1,j+1} + u_{i-1,j})\right\} + \left[\frac{\text{Am An}}{\text{Re Sc}}\right]\delta_2\phi^2 R(u_{i+1/2}) \qquad \textbf{(4.47)}$$

with the requirement that the positions r_{sc} and r_B be included in the set of mesh points r_i. Since $R(u_{i+1/2})$ is a nonlinear function of $u_{i+1/2}$, the extrapolated Crank-Nicolson method (4.37) was used to solve (4.47), and typical results from Davis et al. [7] are shown in Figure 4.6. For a discussion of the physical significance of these results see [7].

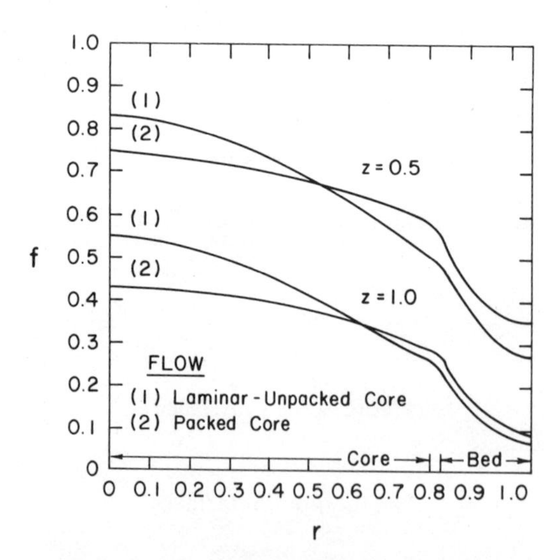

FIGURE 4.6 Results of annular bed reactor. Adapted from M. E. Davis, G. Fairweather, and J. Yamanis, Can. J. Chem. Eng., *59*, p. 499 (1981). Reprinted by permission of the Can. J. Chem. Eng./C.S.Ch.E. and the authors.

High-Order Time Approximations

Selecting an integration method for use in the MOL procedure is very important since it influences accuracy and stability. Until now we have only discussed methods that are $0(\Delta t)$ or $0(\Delta t^2)$. Higher-order integration methods, such as Runge-Kutta, and multistep methods can be used in the MOL procedure, and in fact are the preferred methods in commercial MOL software. The formulation of the MOL-IVP leads to a system of the form:

$$B \frac{d\mathbf{u}}{dt} + A\mathbf{u} = \mathbf{f}$$

$$u_i(0) = \alpha_0(x_i) \tag{4.48}$$

where $\mathbf{u}(t)$ is the vector of nodal solution values at grid points of the mesh (x_i), A corresponds to the spatial operator, $\mathbf{f}$ can be a nonlinear function vector, and $\alpha_0(x_i)$ are the nodal solution values at the grid points for $t = 0$. More generally, (4.48) can be written as:

$$\frac{d\mathbf{u}}{dt} = \mathbf{g}(t, \mathbf{u}) \tag{4.49}$$

It is the eigenvalues of the Jacobian $\boldsymbol{J}$,

$$\boldsymbol{J} = \frac{\partial \mathbf{g}}{\partial \mathbf{u}} \tag{4.50}$$

that determine the stiffness ratio, and thus the appropriate IVP technique.

Sepehrnoori and Carey [8] have examined the effect of the IVP solver on systems of the form (4.48) arising from PDEs by using a nonstiff (ODE [9]), a stiff (DGEAR [10]), and a stabilized Runge-Kutta method (M3RK [11]). Their results confirm that nonstiff, moderately stiff, and stiff systems are most effectively solved by nonstiff, stabilized explicit, and stiff algorithms respectively. Also they observed that systems commonly encountered in practice require both stiff and nonstiff integration algorithms, and that a system may change from stiff to nonstiff or vice versa during the period of integration. Some integrators such as DGEAR offer both stiff and nonstiff options, and the choice is left to the user. This idea merits continued development to include stiffness detection and the ability to switch from one category to another as a significant change in stiffness is detected. Such an extension has been developed by Skeel and Kong [12]. More generalizations of this type will be particularly useful for developing efficient MOL schemes.

Let us examine the MOL using higher-order time approximations by considering the time-dependent mass and energy balances for a porous catalyst particle. Because of the large amount of data required to specify the problem, consider the specific reaction:

$$C_6H_6 \text{ (benzene)} + 3H_2 \text{ (hydrogen)} = C_6H_{12} \text{ (cyclohexane)}$$

which proceeds with the aid of a nickel/kieselguhr catalyst. The material and energy balances for a spherical catalyst pellet are:

$$\varepsilon \frac{\partial C_B}{\partial t} = \frac{D_e}{r^2} \frac{\partial}{\partial r}\left(r^2 \frac{\partial C_B}{\partial r}\right) - R_B \qquad \text{(benzene)}$$

$$\varepsilon \frac{\partial C_H}{\partial t} = \frac{D_e}{r^2} \frac{\partial}{\partial r}\left(r^2 \frac{\partial C_H}{\partial r}\right) - 3R_B \qquad \text{(hydrogen)}$$

$$\rho C_p \frac{\partial T}{\partial t} = \frac{k_e}{r^2} \frac{\partial}{\partial r}\left(r^2 \frac{\partial T}{\partial r}\right) + (-\Delta H)R_B \tag{4.51}$$

with

$$\frac{\partial C_B}{\partial r} = \frac{\partial C_H}{\partial r} = \frac{\partial T}{\partial r} = 0 \quad \text{at} \quad r = 0$$

$$\left.\begin{aligned} D_e \frac{\partial C_B}{\partial r} &= k_g[C_B^0(t) - C_B] \\ D_e \frac{\partial C_H}{\partial r} &= k_g[C_H^0(t) - C_H] \\ k_e \frac{\partial T}{\partial r} &= h_g[T^0(t) - T] \end{aligned}\right\} \text{at} \quad r = 1$$

$$C_B = 0 \quad \text{at} \quad t = 0, \qquad \text{for } 0 \leqslant r \leqslant r_p$$

$$C_H = 0 \quad \text{at} \quad t = 0, \qquad \text{for } 0 \leqslant r \leqslant r_p$$

$$T = T^0(0) \quad \text{at} \quad t = 0, \qquad \text{for } 0 \leqslant r \leqslant r_p$$

where

$-\Delta H$ = heat of reaction
ε = void fraction of the catalyst pellet
C_B, C_H = concentration of benzene, hydrogen
T = temperature
D_e = effective diffusivity (assumed equal for B and H)
k_e = effective thermal conductivity
ρ = density of the fluid-solid system
C_p = heat capacity of the fluid-solid system
r = radial coordinate; r_p = radius of the pellet
t = time
k_g = mass transfer coefficient
h_g = heat transfer coefficient
R_B = reaction rate of benzene, $R_B = R_B(C_B, C_H, T)$

and the superscript 0 represents the ambient conditions. Notice that C_B^0, C_H^0, and T^0 are functions of time; i.e., they can vary with the perturbations in a reactor. Define the following dimensionless parameters

$$
\begin{aligned}
y_{\mathrm{B}} &= \frac{C_{\mathrm{B}}}{C_{\mathrm{B}}^0(0)} \\
y_{\mathrm{H}} &= \frac{C_{\mathrm{H}}}{C_{\mathrm{H}}^0(0)} \\
\theta &= \frac{T}{T^0(0)} \\
x &= \frac{r}{r_p}
\end{aligned} \tag{4.52}
$$

Substitution of these parameters into the transport equations gives:

$$
\begin{aligned}
\frac{\partial y_{\mathrm{B}}}{\partial \tau} &= \frac{1}{x^2}\frac{\partial}{\partial x}\left(x^2 \frac{\partial y_{\mathrm{B}}}{\partial x}\right) - \phi^2 \mathscr{R} \\
\frac{\partial y_{\mathrm{H}}}{\partial \tau} &= \frac{1}{x^2}\frac{\partial}{\partial x}\left(x^2 \frac{\partial y_{\mathrm{H}}}{\partial x}\right) - 3\phi^2 \frac{C_{\mathrm{B}}^0(0)}{C_{\mathrm{H}}^0(0)} \mathscr{R} \\
\mathrm{Le}\,\frac{\partial \theta}{\partial \tau} &= \frac{1}{x^2}\frac{\partial}{\partial x}\left(x^2 \frac{\partial \theta}{\partial x}\right) + \beta\phi^2 \mathscr{R}
\end{aligned} \tag{4.53}
$$

with

$$
\frac{\partial y_{\mathrm{B}}}{\partial x} = \frac{\partial y_{\mathrm{H}}}{\partial x} = \frac{\partial \theta}{\partial x} = 0 \quad \text{at} \quad x = 0
$$

$$
\left.
\begin{aligned}
\frac{\partial y_{\mathrm{B}}}{\partial x} &= \mathrm{Bi}_m \left[\frac{C_{\mathrm{B}}^0(t)}{C_{\mathrm{B}}^0(0)} - y_{\mathrm{B}}\right] \\
\frac{\partial y_{\mathrm{H}}}{\partial x} &= \mathrm{Bi}_m \left[\frac{C_{\mathrm{H}}^0(t)}{C_{\mathrm{H}}^0(0)} - y_{\mathrm{H}}\right] \\
\frac{\partial \theta}{\partial t} &= \mathrm{Bi}_h \left[\frac{T^0(t)}{T^0(0)} - \theta\right]
\end{aligned}
\right\} \quad \text{at} \quad x = 1
$$

$$
\begin{aligned}
y_{\mathrm{B}} &= 0 \quad \text{at} \quad \tau = 0, \qquad \text{for } 0 \leq x \leq 1 \\
y_{\mathrm{H}} &= 0 \quad \text{at} \quad \tau = 0, \qquad \text{for } 0 \leq x \leq 1 \\
\theta &= 1 \quad \text{at} \quad \tau = 0, \qquad \text{for } 0 \leq x \leq 1
\end{aligned}
$$

where

$$\phi^2 = \frac{r_p^2}{C_B^0(0)D_e} R_B(C_B^0(0),\ C_H^0(0),\ T^0(0)) \qquad \text{(Thiele modulus squared)}$$

$$\beta = \frac{D_e(-\Delta H)C_B^0(0)}{k_e T^0(0)} \qquad \text{(Prater number)}$$

$$\text{Le} = \frac{D_e \rho C_p}{k_e \varepsilon} \qquad \text{(Lewis number)}$$

$$\mathcal{R} = \frac{R_B}{R_B(C_B^0(0),\ C_H^0(0),\ T^0(0))}$$

$$\tau = \frac{D_e t}{r_p^2 \varepsilon} \qquad \text{(dimensionless time)}$$

$$\text{Bi}_m = \frac{r_p k_g}{D_e} \qquad \text{(mass Biot number)}$$

$$\text{Bi}_h = \frac{r_p h_g}{k_e} \qquad \text{(heat Biot number)}$$

For the benzene hydrogenation reaction, the reaction rate function is [13]:

$$R_B = \frac{\rho_{\text{cat}} k\, K \exp\left[\dfrac{(Q - E)}{R_g T}\right] P_B P_H}{1 + K \exp\left(\dfrac{Q}{R_g T}\right) P_B} \tag{4.54}$$

where

$$\begin{aligned}
k &= 3207 \text{ gmole/(sec·atm·gcat)} \\
K &= 3.207 \times 10^{-8} \text{ atm}^{-1} \\
Q &= 16{,}470 \text{ cal/gmole} \\
E &= 13{,}770 \text{ cal/gmole} \\
R_g &= 1.9872 \text{ cal/(gmole·K)} \\
\rho_{\text{cat}} &= 1.88 \text{ g/cm}^3 \\
P_i &= \text{partial pressure of component } i
\end{aligned}$$

Noticing that

$$y_i = \frac{C_i}{C_i^0(0)} = \frac{P_i}{P_i^0(0)}\left(\frac{T^0(0)}{T}\right)$$

$$\mathcal{R} = \exp\left[\alpha_2\left(\frac{1}{\theta} - 1\right)\right]\theta^2 y_B y_H \frac{[1 + KP_B^0(0)\exp(\alpha_1)]}{\left[1 + KP_B^0(0)\exp\left(\dfrac{\alpha_1}{\theta}\right) y_B \theta\right]} \tag{4.55}$$

where

$$\alpha_1 = \frac{Q}{R_g\, T^0(0)}$$

$$\alpha_2 = \frac{(Q - E)}{R_g\, T^0(0)}$$

We solve this system of equations using the MOL. First, the system is written as an IVP by discretizing the spatial derivatives using difference formulas. The IVP is:

$$\frac{\partial y_{B,1}}{\partial \tau} = \frac{6}{h^2}[y_{B,2} - y_{B,1}] - \phi^2 \mathscr{R}_1 \qquad \text{(use false boundaries)}$$

$$\frac{\partial y_{H,1}}{\partial \tau} = \frac{6}{h^2}[y_{H,2} - y_{H,1}] - 3\phi^2 \frac{C_B^0(0)}{C_H^0(0)} \mathscr{R}_1$$

$$\text{Le}\,\frac{\partial \theta_1}{\partial \tau} = \frac{6}{h^2}[\theta_2 - \theta_1] + \beta\phi^2 \mathscr{R}_1$$

$$\frac{\partial y_{B,i}}{\partial \tau} = \frac{1}{h^2}\left\{\left[1 + \frac{1}{(i-1)}\right]y_{B,i+1} - 2y_{B,i} + \left[1 - \frac{1}{(i-1)}\right]y_{B,i-1}\right\} - \phi^2 \mathscr{R}_i$$

$$\frac{\partial y_{H,i}}{\partial \tau} = \frac{1}{h^2}\left\{\left[1 + \frac{1}{(i-1)}\right]y_{H,i+1} - 2y_{H,i} + \left[1 - \frac{1}{(i-1)}\right] y_{H,i-1}\right\} - 3\phi^2 \frac{C_B^0(0)}{C_H^0(0)} \mathscr{R}_i$$

$$\text{Le}\,\frac{\partial \theta_i}{\partial \tau} = \frac{1}{h^2}\left\{\left[1 + \frac{1}{(i-1)}\right]\theta_{i+1} - 2\theta_i + \left[1 - \frac{1}{(i-1)}\right]\theta_{i-1}\right\} + \beta\phi^2 \mathscr{R}_i \qquad \textbf{(4.56)}$$

with $i = 2, \ldots, N$,

$$\frac{\partial y_{B,N+1}}{\partial \tau} = \frac{1}{h^2}\left\{2y_{B,N} - 2\left[1 + \left(1 + \frac{1}{N}\right)\text{Bi}_m h\right]y_{B,N+1} + 2\text{Bi}_m h \frac{C_B^0(t)}{C_B^0(0)}\left[1 + \frac{1}{N}\right]\right\} - \phi^2 \mathscr{R}_{N+1}$$

$$\frac{\partial y_{H,N+1}}{\partial \tau} = \frac{1}{h^2}\left\{2y_{H,N} - 2\left[1 + \left(1 + \frac{1}{N}\right)\text{Bi}_m h\right]y_{H,N+1} + 2\text{Bi}_m h \frac{C_H^0(t)}{C_H^0(0)}\left[1 + \frac{1}{N}\right]\right\} - 3\phi^2 \frac{C_B^0(0)}{C_H^0(0)} \mathscr{R}_{N+1}$$

$$\mathrm{Le}\,\frac{\partial\theta_{N+1}}{\partial\tau} = \frac{1}{h^2}\left\{2\theta_N - 2\left[1 + \left(1 + \frac{1}{N}\right)\mathrm{Bi}_h h\right]\theta_{N+1} + 2\mathrm{Bi}_h h\,\frac{T^0(t)}{T^0(0)}\left[1 + \frac{1}{N}\right]\right\} + \beta\phi^2\,\mathcal{R}_{N+1}$$

where

$$h = \Delta x$$
$$y_{\mathrm{B},i} \simeq y_{\mathrm{B}}(x_i)$$
$$\mathcal{R}_i = \mathcal{R}(y_{\mathrm{B},i}, y_{\mathrm{H},i}, \theta_i)$$

Notice that the Jacobian of the right-hand-side vector of the above system has a banded structure. That is, if J_{ij} are the elements of the Jacobian, then

$$J_{ij} = 0, \qquad \text{for } |i - j| \geq 4$$

This system is integrated using the software package LSODE [14] (see Chapter 1) since this routine contains nonstiff and stiff multistep integration algorithms (Gear's method) and also allows for Jacobians that possess banded structure. The data used in the following simulations are listed in Table 4.2.

The physical situation can be explained as follows. Initially the catalyst pellet is bathed in an inert fluid. At $t = 0$ a reaction mixture composed of 2.5% benzene and the remainder hydrogen is forced past the catalyst pellet. The dimensionless benzene profiles are shown in Figure 4.7. With increasing time (τ), the benzene is able to diffuse further into the pellet. If no reaction occurs, the profile at large τ would be the horizontal line at $y_{\mathrm{B}} = 1.0$. The steady-state profile ($\tau \to \infty$) is not significantly different than the one shown at $\tau = 1.0$.

A MOL algorithm contains two portions that produce errors: the spatial discretization and the time integration. In the following results the time integration error is controlled by the parameter TOL, while the spatial discretization error is a function of h, the step-size. For the results shown, TOL $= 10^{-5}$. A decrease in the value of TOL did not affect the results (to the number of significant figures shown). The effects of h are shown in Table 4.3. (The results of specifying $h_4 = h_3/2$ were the same as those shown for h_3.) Therefore, as h decreases, the spatial error is decreased. We would expect the spatial error to be $0(h^2)$ since a second-order correct finite difference discretization is used.

TABLE 4.2 Parameter Data for Catalyst Start-up Simulation†

$\phi = 1.0$	$C_{\mathrm{B}}^0(0)/C_{\mathrm{H}}^0 = 0.025/0.975$
$\beta = 0.04$	$C_{\mathrm{B}}^0(t) = C_{\mathrm{B}}^0(0), \quad t \geq 0$
$\mathrm{Le} = 80$	$C_{\mathrm{H}}^0(t) = C_{\mathrm{H}}^0(0), \quad t \geq 0$
$\mathrm{Bi}_m = 350$	$T^0(t) = T^0(0), \quad t \geq 0$
$\mathrm{Bi}_h = 20$	$T^0(0) = 373.15$ K

† From reference [15].

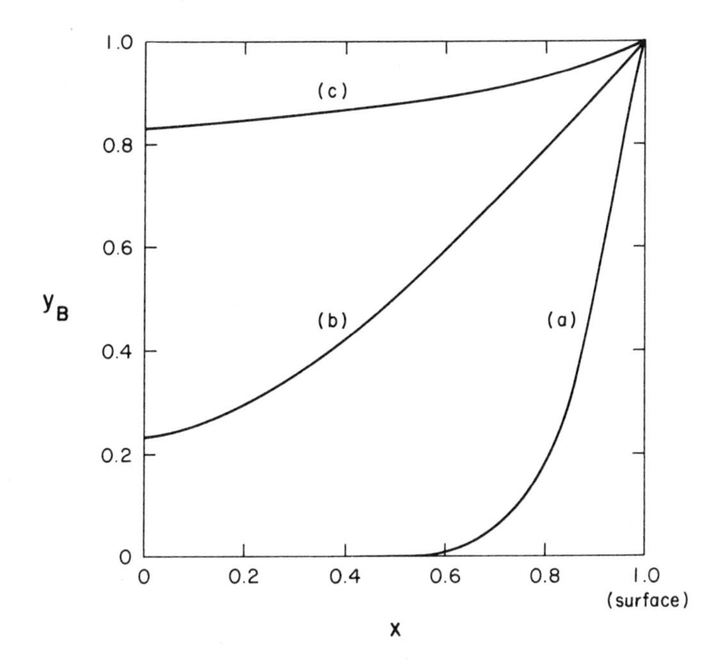

FIGURE 4.7 Solution of Eq. (4.51)

	τ
(*a*)	0.01
(*b*)	0.10
(*c*)	1.00

The effects of the method of time integration are shown in Table 4.4. Notice that this problem is stiff since the nonstiff integration required much more execution time for solution. From Table 4.4 one can see that using the banded Jacobian feature cut the execution time in half over that when using the full Jacobian option. This is due to less computations during matrix factorizations, matrix multiplications, etc.

TABLE 4.3 Results of Catalyst Start-Up
LSODE TOL = 10^{-5}
Benzene Profile, y_B, at τ = 0.1

x	$h_1 = 0.125$	$h_2 = \frac{h_1}{2}$	$h_3 = \frac{h_2}{2}$
0.00	0.2824	0.2773	0.2760
0.25	0.3379	0.3341	0.3331
0.50	0.4998	0.4987	0.4983
0.75	0.7405	0.7408	0.7408
1.00	0.9973	0.9973	0.9973

TABLE 4.4 Further Results of Catalyst Start-Up
LSODE TOL = 10^{-5}
h = 0.0625

	Execution Time Ratio	
	$\tau = 0.1$	$\tau = 1.0$
Nonstiff method	7.87	50.98
Stiff method (full Jacobian)	2.30	2.24
Stiff method (banded Jacobian)	1.00	1.00

FINITE ELEMENTS

In the following sections we will continue our discussion of the MOL, but will use finite element discretizations in the space variable. As in Chapter 3, we will first present the Galerkin method, and then proceed to outline collocation.

Galerkin

Here, the Galerkin methods outlined in Chapter 3 are extended to parabolic PDEs. To solve a PDE, specify the piecewise polynomial approximation (pp-approximation) to be of the form:

$$u(x, t) = \sum_{j=1}^{m} \alpha_j(t)\phi_j(x) \tag{4.57}$$

Notice that the coefficients are now functions of time. Consider the PDE:

$$\frac{\partial w}{\partial t} = \frac{\partial}{\partial x}\left[\hat{a}(x, w)\,\frac{\partial w}{\partial x}\right], \qquad 0 < x < 1, \quad t > 0 \tag{4.58}$$

with

$$w(x, 0) = w_0(x)$$
$$w(0, t) = 0$$
$$w(1, t) = 0$$

If the basis functions are chosen to satisfy the boundry conditions, then the weak form of (4.58) is

$$\left(\hat{a}(x, w)\,\frac{\partial w}{\partial x}, \phi_i'\right) + \left(\frac{\partial w}{\partial t}, \phi_i\right) = 0 \qquad i = 1, \ldots m \tag{4.59}$$

The MOL formulation for (4.58) is accomplished by substituting (4.57) into (4.59) to give

$$\left(\hat{a}\left[x, \sum_{j=1}^{m} \alpha_j \phi_j\right] \sum_{j=1}^{m} \alpha_j \phi_j', \phi_i'\right) + \left(\sum_{j=1}^{m} \alpha_j' \phi_j, \phi_i\right) = 0 \qquad \textbf{(4.60)}$$

$$i = 1, \ldots, m$$

Thus, the IVP (4.60) can be written as:

$$\boldsymbol{B}\boldsymbol{\alpha}'(t) + \boldsymbol{D}(\boldsymbol{\alpha})\boldsymbol{\alpha}(t) = \mathbf{0} \qquad \textbf{(4.61)}$$

with

$$\boldsymbol{B}\boldsymbol{\alpha}(0) = \boldsymbol{\beta}$$

where

$$D_{i,j} = (\hat{a}\phi_j', \phi_i')$$

$$B_{i,j} = (\phi_j, \phi_i)$$

$$\boldsymbol{\beta} = [(w_0, \phi_1), (w_0, \phi_2), \ldots, (w_0, \phi_m)]^T$$

In general this problem is nonlinear as a result of the function $\hat{a}$ in the matrix $\boldsymbol{D}$. One could now use an IVP solver to calculate $\boldsymbol{\alpha}(t)$. For example, if the trapezoidal rule is applied to (4.61), the result is:

$$\boldsymbol{B}\left[\frac{\boldsymbol{\alpha}^{n+1} - \boldsymbol{\alpha}^n}{\Delta t}\right] + \boldsymbol{D}(\boldsymbol{\alpha}^{n+1/2})\boldsymbol{\alpha}^{n+1/2} = \mathbf{0} \qquad \textbf{(4.62)}$$

with

$$\boldsymbol{B}\boldsymbol{\alpha}^0 = \boldsymbol{\beta}$$

where

$$t_n = n\,\Delta t$$

$$\boldsymbol{\alpha}^n \simeq \boldsymbol{\alpha}(t_n)$$

$$\boldsymbol{\alpha}_{n+1/2} = \frac{\boldsymbol{\alpha}^{n+1} + \boldsymbol{\alpha}^n}{2}$$

Since the basis ϕ_i is local, the matrices $\boldsymbol{B}$ and $\boldsymbol{D}$ are sparse. Also, recall that the error in the Galerkin discretization is $0(h^k)$ where k is the order of the approximating space. Thus, for (4.62), the error is $0(\Delta t^2 + h^k)$, since the trapezoidal rule is second-order accurate. In general, the error of a Galerkin-MOL discretization is $0(\Delta t^p + h^k)$ where p is determined by the IVP solver. In the following example we extend the method to a system of two coupled equations.

EXAMPLE 4

A two-dimensional model is frequently used to describe fixed-bed reactors that are accomplishing highly exothermic reactions. For a single independent reaction a material and energy balance is sufficient to describe the reactor. These continuity equations are:

$$r\frac{\partial f}{\partial z} = \text{Pe}\left[\frac{\partial}{\partial r}\left(r\frac{\partial f}{\partial r}\right)\right] + \beta_1 r\mathscr{R}(f, \theta)$$

$$r\frac{\partial \theta}{\partial z} = \text{Bo}\left[\frac{\partial}{\partial r}\left(r\frac{\partial \theta}{\partial r}\right)\right] + \beta_2 r\mathscr{R}(f, \theta)$$

with

$$f = \theta = 1 \quad \text{at} \quad z = 1, \qquad \text{for } 0 < r < 1$$

$$\frac{\partial f}{\partial r} = \frac{\partial \theta}{\partial r} = 0 \quad \text{at} \quad r = 0, \qquad \text{for } 0 < z < 1$$

$$\frac{\partial f}{\partial r} = 0, \qquad \frac{\partial \theta}{\partial r} = -\text{Bi}(\theta - \theta_w) \quad \text{at} \quad r = 1, \qquad \text{for } 0 < z < 1$$

where

z = dimensionless axial coordinate, $0 \leqslant z \leqslant 1$
r = dimensionless radial coordinate, $0 \leqslant r \leqslant 1$
f = dimensionless concentration
θ = dimensionless temperature
θ_w = dimensionless reactor wall temperature
$\mathscr{R}$ = dimensionless reaction rate function
Bi, Pe, Bo, β_1, β_2 = constants

These equations express the fact that the change in the convection is equal to the change in the radial dispersion plus the change due to reaction. The boundary condition for θ at $r = 1$ comes from the continuity of heat flux at the reactor wall (which is maintained at a constant value of θ_w). Notice that these equations are parabolic PDE's, and that they are nonlinear and coupled through the reaction rate function. Set up the Galerkin-MOL-IVP.

SOLUTION

Let

$$u(x, z) = \sum_{j=1}^{m} \alpha_j(z)\phi_j(x) \simeq f(x, z)$$

$$v(x, z) = \sum_{j=1}^{m} \gamma_j(z)\phi_j(x) \simeq \theta(x, z)$$

such that

$$u(x, 0) = \sum_{j=1}^{m} \alpha_j(0)\phi_j(x) = 1.0$$

$$v(x, 0) = \sum_{j=1}^{m} \gamma_j(0)\phi_j(x) = 1.0$$

The weak form of this problem is

$$\left(\frac{\partial f}{\partial z}, \phi_i\right) = \text{Pe}\left[r\frac{\partial f}{\partial r}\phi_i\bigg|_0^1 - \left(\frac{\partial f}{\partial r}, \phi_i'\right)\right] + \beta_1(\mathcal{R}, \phi_i)$$

$$\left(\frac{\partial \theta}{\partial z}, \phi_i\right) = \text{Bo}\left[r\frac{\partial \theta}{\partial r}\phi_i\bigg|_0^1 - \left(\frac{\partial \theta}{\partial r}, \phi_i'\right)\right] + \beta_2(\mathcal{R}, \phi_i)$$

$$\text{i} = 1, \ldots, m$$

where

$$(a, b) = \int_0^1 abr\, dr$$

The boundary conditions specify that

$$r\frac{\partial f}{\partial r}\phi_i\bigg|_0^1 = 0 \qquad \left(\frac{\partial f}{\partial r} = 0 \quad \text{at } r = 1 \text{ and } r = 0\right)$$

$$r\frac{\partial \theta}{\partial r}\phi_i\bigg|_0^1 = -\text{Bi}\,(\theta - \theta_w)\phi_i(1) \qquad \left(\frac{\partial \theta}{\partial r} = 0 \quad \text{at} \quad r = 0\right)$$

Next, substitute the pp-approximations for f and θ into the weak forms of the continuity equations to give

$$\sum_{j=1}^{m} \alpha_j'(z)(\phi_j, \phi_i) = -\text{Pe}\left[\sum_{j=1}^{m} \alpha_j(z)(\phi_j', \phi_i')\right] + \beta_1(\tilde{\mathcal{R}}, \phi_i)$$

$$\sum_{j=1}^{m} \gamma_j'(z)(\phi_j, \phi_i) = -\text{Bo}\left[\text{Bi}\left(\sum_{j=1}^{m} \gamma_j\phi_j(1) - \theta_w\right)\phi_i(1)\right.$$

$$\left. + \sum_{j=1}^{m} \gamma_j(z)(\phi_j', \phi_i')\right] + \beta_2(\tilde{\mathcal{R}}, \phi_i), \qquad \text{i} = 1, \ldots, m$$

where

$$\tilde{\mathcal{R}} = \mathcal{R}\left(\sum_{j=1}^{m} \alpha_j\phi_j, \sum_{j=1}^{m} \gamma_i\phi_j\right)$$

Denote

$$
\begin{aligned}
C_{i,j} &= (\phi_j, \phi_i) \\
D_{i,j} &= (\phi_j', \phi_j') \\
B_{i,j} &= \phi_i(1)\phi_j(1) \\
\psi_i &= (\tilde{\mathscr{R}}, \phi_i)
\end{aligned}
$$

The weak forms of the continuity equations can now be written as:

$$
\begin{aligned}
\boldsymbol{C}\boldsymbol{\alpha}' &= -\text{Pe}\,\boldsymbol{D}\boldsymbol{\alpha} + \beta_1\boldsymbol{\psi} \\
\boldsymbol{C}\boldsymbol{\gamma}' &= -\text{Bo}\,[\text{Bi}\,(\boldsymbol{B}\boldsymbol{\gamma} - \theta_w\boldsymbol{\phi}(1)) + \boldsymbol{D}\boldsymbol{\gamma}] + \beta_2\boldsymbol{\psi}
\end{aligned}
$$

with $\boldsymbol{\alpha}(0)$ and $\boldsymbol{\gamma}(0)$ known. Notice that this system is a set of IVPs in the dependent variables $\boldsymbol{\alpha}$ and $\boldsymbol{\gamma}$ with initial conditions $\boldsymbol{\alpha}(0)$ and $\boldsymbol{\gamma}(0)$. Since the basis ϕ_i is local, the matrices $\boldsymbol{B}$, $\boldsymbol{C}$, and $\boldsymbol{D}$ are sparse. The solution of this system gives $\boldsymbol{\alpha}(t)$ and $\boldsymbol{\gamma}(t)$, which completes the specification of the pp-approximations u and v.

From the foregoing discussions one can see that there are two components to any Galerkin-MOL solution, namely the spatial discretization procedure and the IVP solver. We defer further discussions of these topics and that of Galerkin-based software until the end of the chapter where it is covered in the section on mathematical software.

Collocation

As in Chapter 3 we limit out discussion to spline collocation at Gaussian points. To begin, specify the pp-approximation to be (4.57), and consider the PDE:

$$
\frac{\partial w}{\partial t} = f\left(x, t, w, \frac{\partial w}{\partial x}, \frac{\partial^2 w}{\partial x^2}\right), \qquad 0 < x < 1, \quad t > 0 \tag{4.63}
$$

with

$$
\begin{aligned}
w &= w_0(x) \quad \text{at} \quad t = 0 \\
\eta_1 w + \beta_1 w' &= \gamma_1(t) \quad \text{at} \quad x = 0 \\
\eta_2 w + \beta_2 w' &= \gamma_2(t) \quad \text{at} \quad x = 1
\end{aligned}
$$

where

$$
|\eta_i| + |\beta_i| > 0, \qquad i = 1, 2
$$

According to the method of spline collocation at Gaussian points, we require that (4.57) satisfy the differential equation at $(k - M)$ Gaussian points per subinterval, where k is the degree of the approximating space and M is the order of the differential equation. Thus with the mesh

$$
0 = x_1 < x_2 < \ldots < x_{\ell+1} = 1 \tag{4.64}
$$

Eq. (4.63) becomes

$$\sum_{j=1}^{m} \alpha_j'(t)\phi_j(x_i) = f\left(x_i, t, \sum_{j=1}^{m} \alpha_j(t)\phi_j(x_i), \sum_{j=1}^{m} \alpha_j(t)\phi_j'(x_i), \sum_{j=1}^{m} \alpha_j(t)\phi''(x_i)\right)$$

$$i = 1, \ldots, m - 2 \tag{4.65}$$

where

$$m = (k - M)\ell + 2$$
$$M = 2$$

We now have $m - 2$ equations in m unknown coefficients α_j. The last two equations necessary to specify the pp-approximation are obtained by differentiating the boundary conditions:

$$\eta_1 \sum_{j=1}^{m} \alpha_j'(t)\phi_j(0) + \beta_1 \sum_{j=1}^{m} \alpha_j'(t)\phi_j'(0) = \gamma_1'(t)$$

$$\eta_2 \sum_{j=1}^{m} \alpha_j'(t)\phi_j(1) + \beta_2 \sum_{j=1}^{m} \alpha_j'(t)\phi_j'(1) = \gamma_2'(t) \tag{4.66}$$

This system of equations can be written as

$$\boldsymbol{A\alpha}'(t) = \mathbf{F}(t, \boldsymbol{\alpha})$$

$$\boldsymbol{\alpha}(0) = \boldsymbol{\alpha}_0 \tag{4.67}$$

where

$$\boldsymbol{A} = \text{left-hand side of (4.65) and (4.66)}$$
$$\boldsymbol{F} = \text{right-hand side of (4.65) and (4.66)}$$

and $\boldsymbol{\alpha}_0$ is given by:

$$\sum_{j=1}^{m} \alpha_j(0)\phi_j(x_i) = w_0(x_i), \qquad i = 1, \ldots, m$$

Since the basis ϕ_j is local, the matrix $\boldsymbol{A}$ will be sparse. Equation (4.67) can now be integrated in time by an IVP solver. As with Galerkin-MOL, the error produced from collocation-MOL is $0(\Delta t^p + h^k)$ where p is specified by the IVP solver and k is set by the choice of the approximating space. In the following example we formulate the collocation method for a system of two coupled equations.

EXAMPLE 5

A polymer solution is spun into an acidic bath, where the diffusion of acid into the polymeric fiber causes the polymer to coagulate. We wish to find the concentration of the acid (in the fiber), C_A, along the spinning path. The coagulation reaction is

$$\text{Polymer} + \text{acid} \rightarrow \text{coagulated polymer} + \text{salt}$$

The acid bath is well stirred with the result that the acid concentration at the surface of the fiber is constant. The governing equations are the material balance equations for the acid and the salt within the polymeric fiber, and are given by

$$u \frac{\partial C_A}{\partial z} = \frac{1}{r} \frac{\partial}{\partial r} \left(r D_A(C_s) \frac{\partial C_A}{\partial r} \right) - k C_A, \qquad 0 < r < r_f,\ z > 0$$

$$u \frac{\partial C_s}{\partial z} = \frac{1}{r} \frac{\partial}{\partial r} \left(r D_s(C_s) \frac{\partial C_s}{\partial r} \right) + k C_A$$

where

C_s = concentration of the salt

z = axial coordinate

r = radial coordinate

r_f = radius of the fiber

u = axial velocity of the fiber as it moves through the acid bath

D_A = acid diffusivity in the fiber

D_s = salt diffusivity in the fiber

k = first-order coagulation constant

The subsidiary conditions are

$C_A = 0$ at $z = 0$ (no acid initially present)

$\frac{\partial C_s}{\partial r} = \frac{\partial C_A}{\partial r} = 0$ at $r = 0$ (symmetry)

$C_A = C_A^0$ at $r = r_f$ (uniform concentration at fiber–acid bath interface)

$C_s = 0$ at $z = 0$ (no salt initially present)

$C_s = 0$ at $r = r_f$ (salt concentration of fiber–acid bath interface maintained at zero by allowing the acid bath to be an infinite sink for the salt)

Let

$$D_A(C_s) = D_0 e^{-\eta C_s}$$

$$D_s(C_s) = D_0 e^{-\lambda C_s}$$

where D_0, λ, and η are constants. Set up the collocation-MOL-IVP with $k = 4$.

SOLUTION

Let

$$u(r, z) = \sum_{j=1}^{m} \alpha_j(z)\phi_j(r) \simeq C_A(r, z)$$

$$v(r, z) = \sum_{j=1}^{m} \gamma_j(z)\phi_j(r) \simeq C_s(r, z)$$

such that

$$u(r, 0) = \sum_{j=1}^{m} \alpha_j(0)\phi_j(r) = 0$$

$$v(r, 0) = \sum_{j=1}^{m} \gamma_j(0)\phi_j(r) = 0$$

for

$$0 = x_1 < x_2 < \ldots < x_\ell < x_{\ell+1} = r_f$$

Since $k = 4$ and $M = 2$, $m = 2(\ell + 1)$ and there are two collocation points per subinterval, τ_{i1} and τ_{i2}, $i = 1, \ldots, \ell$. For a given subinterval i, the collocation equations are

$$u \sum_{j=1}^{m} \alpha_j'(z)\phi_j(\tau_{is}) = \overline{D}_A(\tau_{is}) \left[\sum_{j=1}^{m} \alpha_j(z)\phi_j''(\tau_{is}) + \frac{1}{\tau_{is}} \sum_{j=1}^{m} \alpha_j(z)\phi_j'(\tau_{is}) - \eta \left(\sum_{j=1}^{m} \gamma_j(z)\phi_j'(\tau_{is}) \right) \left(\sum_{j=1}^{m} \alpha_j(z)\phi_j'(\tau_{is}) \right) \right] - k \sum_{j=1}^{m} \alpha_j(z)\phi_j(\tau_{is})$$

$$u \sum_{j=1}^{m} \gamma_j'(z)\phi_j(\tau_{is}) = \overline{D}_s(\tau_{is}) \left[\sum_{j=1}^{m} \gamma_j(z)\phi_j''(\tau_{is}) + \frac{1}{\tau_{is}} \sum_{j=1}^{m} \gamma_j(z)\phi_j'(\tau_{is}) - \lambda \left(\sum_{j=1}^{m} \alpha_j(z)\phi_j'(\tau_{is}) \right)^2 \right] + k \sum_{j=1}^{m} \alpha_j(z)\phi_j(\tau_{is})$$

for $s = 1, 2$, where

$$\overline{D}_A(\tau_{is}) = D_0 \exp\left[-\eta \left(\sum_{j=1}^{m} \gamma_j(z)\phi_j(\tau_{is}) \right) \right]$$

$$\overline{D}_s(\tau_{is}) = D_0 \exp\left[-\lambda \left(\sum_{j=1}^{m} \gamma_j(z)\phi_j(\tau_{is}) \right) \right]$$

At the boundaries we have the following:

$$\sum_{j=1}^{m} \alpha_j'(z)\phi_j'(0) = \sum_{j=1}^{m} \gamma_j'(z)\phi_j'(0) = 0$$

$$\sum_{j=1}^{m} \alpha_j'(z)\phi_j(r_f) = C_A^0, \qquad \sum_{j=1}^{m} \gamma_j'(z)\phi_j(r_f) = 0$$

The 4ℓ equations at the collocation points and the four boundary equations give $4(\ell + 1)$ equations for the $4(\ell + 1)$ unknowns $\alpha_j(z)$ and $\gamma_j(z)$, $j = 1, \ldots, 2(\ell + 1)$. This system of equations can be written as

$$\begin{aligned} \boldsymbol{A}\boldsymbol{\psi}' &= \mathbf{F}(\boldsymbol{\alpha}(z), \boldsymbol{\gamma}(z)) \\ \boldsymbol{\psi}(0) &= \mathbf{Q} \end{aligned}$$

where

$$\begin{aligned}
\mathbf{Q} &= [0, \ldots, 0, C_A^0, 0]^T \\
A &= \text{sparse matrix} \\
\mathbf{F}(\boldsymbol{\alpha}(z), \boldsymbol{\gamma}(z)) &= \text{nonlinear vector} \\
\boldsymbol{\alpha}(z) &= [\alpha_1(z), \ldots, \alpha_m(z)]^T \\
\boldsymbol{\gamma}(z) &= [\gamma_1(z), \ldots, \gamma_m(z)]^T \\
\boldsymbol{\psi}(z) &= [\alpha_1(z), \gamma_1(z), \alpha_2(z), \gamma_2(z), \ldots, \alpha_m(z), \gamma_m(z)]^T
\end{aligned}$$

The solution of the IVP gives $\boldsymbol{\alpha}(t)$ and $\boldsymbol{\gamma}(t)$, which completes the specifications of the pp-approximations u and v.

As with Galerkin-MOL, a collocation-MOL code must address the problems of the spatial discretization and the "time" integration. In the following section we discuss these problems.

MATHEMATICAL SOFTWARE

A computer algorithm based upon the MOL must include two portions: the spatial discretization routine and the time integrator. If finite differences are used in the spatial discretization procedure, the IVP has the form shown in (4.49), which is that required by the IVP software discussed in Chapter 1. A MOL algorithm that uses a finite element method for the spatial discretization will produce an IVP of the form:

$$A(\mathbf{y}, t)\mathbf{y}' = \mathbf{g}(\mathbf{y}, t) \tag{4.68}$$

Therefore, in a finite element MOL code, implementation of the IVP software discussed in Chapter 1 required that the implicit IVP [Eq. (4.68)] be converted to its explicit form [Eq. (4.49)]. For example, (4.68) can be written as

$$\mathbf{y}' = A^{-1}\mathbf{g} \tag{4.69}$$

where

$$A^{-1} = \text{inverse of } A$$

This problem can be avoided by using the IVP solver GEARIB [16] or its update LSODI [14], which allows for the IVP to be (4.68) where A and $\partial\mathbf{g}/\partial\mathbf{y}$ are banded, and are the implicit forms of the GEAR/GEARB [17,18] or LSODE [14] codes.

Table 4.5 lists the parabolic PDE software and outlines the type of spatial discretization and time integration for each code. Notice that each of the major libraries—NAG, Harwell, and IMSL—contain PDE software. As is evident from Table 4.5, the overwhelming choice of the time integrator is the Gear method. This method allows for stiff and nonstiff equations and has proven reliable over recent years (see users guides to GEAR [16], GEARB [17], and LSODE [14]). Since we have covered the MOL using finite differences in greater detail than

TABLE 4.5 Parabolic PDE Codes

Name	Spatial Discretization	Time Integrator	Reference
NAG (D03 Chapter)	Finite differences	Gear's method, i.e., Adams multistep or implicit multistep	[19]
Harwell (DP01, DP02)	Finite differences	Trapezoidal rule	[20]
IMSL (DPDES)	Collocation; Hermite cubic basis	Gear's method (DGEAR)	[21]
PDEPACK	Finite differences	Gear's method (GEARB)	[22]
DSS/2	Finite differences	Several including Runge-Kutta and Gear's method (GEARB)	[23]
MOL1D	Finite differences	Gear's method (GEARB)	[24]
PDECOL	Collocation; B-spline basis	Gear's method (GEARIB)	[25]
DISPL	Galerkin; B-spline basis	Gear's method (GEARIB)	[26]
POST (in PORT library [28])	Galerkin; B-spline basis	Explicit or implicit one-step with extrapolation	[27]
FORSIM	Finite differences	Several including Runge-Kutta and Gear's method	[29]

when using finite elements, we will finish the discussion of software by solving a PDE with a collocation (PDECOL) and a Galerkin (DISPL) code.

Consider the problem of desorption of a gas from a liquid stream in a wetted wall column. The situation is shown in Figure 4.8*a*. A saturated liquid enters the top of the column and flows downward where it is contacted with a stream of gas flowing countercurrently. This gas is void of the species being desorbed from the liquid. If the radius of the column R_c is large compared with the thickness of the liquid film, then the problem can be solved in rectangular coordinates as shown in Figure 4.8*b*. The material balance of the desorbing species within the liquid film is:

$$u\left[1 - \left(\frac{x}{x_f}\right)^2\right]\frac{\partial C}{\partial z} = D\frac{\partial^2 C}{\partial x^2} \tag{4.70}$$

with

$$\frac{\partial C}{\partial x} = 0 \quad \text{at} \quad x = x_f$$

$$C = 0 \quad \text{at} \quad x = 0$$

$$C = C^* \quad \text{at} \quad z = 0$$

where

C = concentration of the desorbing species in the liquid film

C^* = saturation concentration of the desorbing species

u = maximum liquid velocity

D = diffusivity of the desorbing species

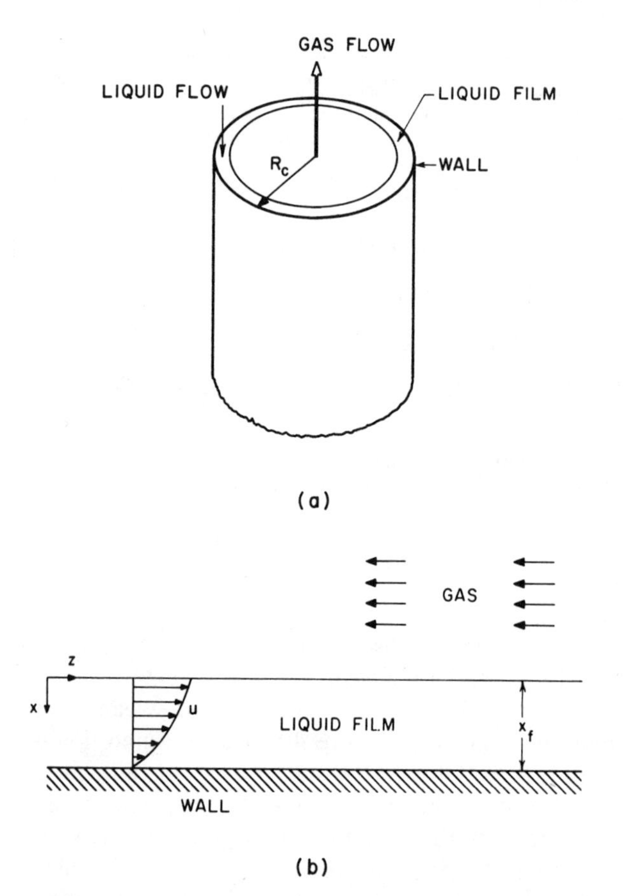

FIGURE 4.8 Desorption in a wetted-wall column. (*a*) Macroscopic schematic. (*b*) Microscopic schematic.

The boundary condition at $x = 0$ implies that there is no gas film transport resistance and that the bulk gas concentration of the desorbing species is zero. Let

$$f = \frac{C}{C^*}$$

$$\eta = \frac{x}{x_f}$$

$$\xi = \frac{zD}{ux_f^2} \tag{4.71}$$

Substituting (4.71) into (4.70) gives:

$$(1 - \eta^2)\frac{\partial f}{\partial \xi} = \frac{\partial^2 f}{\partial \eta^2}$$

$$\frac{\partial f}{\partial \eta} = 0 \quad \text{at} \quad \eta = 1$$

$$f = 0 \quad \text{at} \quad \eta = 0$$

$$f = 1 \quad \text{at} \quad z = 0 \tag{4.72}$$

Although (4.72) is linear, it is still somewhat of a "difficult" problem because of the steep gradient in the solution near $\eta = 0$ (see Figure 4.9).

The results for (4.72) using PDECOL and DISPL are shown in Table 4.6. The order of the approximating space, k, was fixed at 4, and the tolerance on the time integration, TOL, set at 10^{-6}. Neither PDECOL or DISPL produced a solution close to those shown in Table 4.6 when using a uniform mesh with $\eta_i - \eta_{i-1} = 0.1$ for all i. Thus the partition was graded in the region of [0, 0.1]

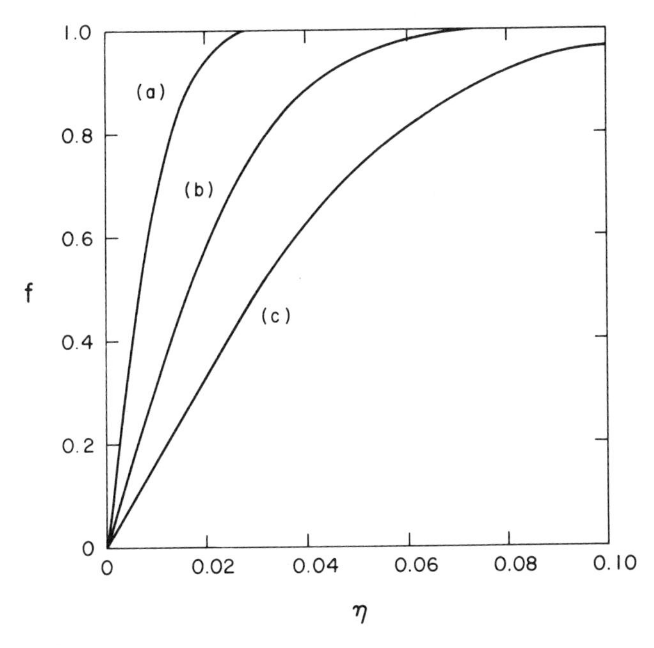

FIGURE 4.9 Solution of Eq. (4.72)

ξ

(*a*) 5×10^{-5}
(*b*) 3×10^{-4}
(*c*) 1×10^{-3}

TABLE 4.6 Results of Eq. (4.72)
$\eta = 0.01$, $k = 4$, TOL = (-6)

	PDECOL		DISPL	
ξ	π_1	π_2	π_1	π_2
5(−5)	0.6827	0.6828	0.6839	0.6828
3(−4)	0.3169	0.3168	0.3169	0.3168
1(−3)	0.1768	0.1769	0.1769	0.1769
π_1:	$0 = \eta_1 < \eta_2 < \ldots \eta_{11} = 0.1$		$\eta_i - \eta_{i-1} = 0.01$	
	$0.1 = \eta_{11} < \eta_{12} < \ldots < \eta_{20} = 1.0$		$\eta_i - \eta_{i-1} = 0.1$	
π_2:	$0 = \eta_1 < \eta_2 < \ldots < \eta_{21} = 0.1$		$\eta_i - \eta_{i-1} = 0.005$	
	$0.1 = \eta_{21} < \eta_{22} < \ldots \eta_{30} = 1.0$		$\eta_i - \eta_{i-1} = 0.1$	

as specified by π_1 and π_2 in Table 4.6. Further mesh refinements in the region of [0, 0.1] did not cause the solutions to differ from those shown for π_2. From these results, one can see that PDECOL and DISPL produced the same solutions. This is an expected result since both codes used the same approximating space, and the same tolerance for the time integration.

The parameter of engineering significance in this problem is the Sherwood number, Sh, which is defined as

$$\mathrm{Sh} = \frac{\left.\left(\dfrac{\partial f}{\partial \eta}\right)\right|_{\eta = 0}}{\bar{f}} \tag{4.73}$$

where

$$\bar{f} = \frac{3}{2}\int_0^1 (1 - \eta^2) f \, d\eta \tag{4.74}$$

Table 4.7 shows the Sherwood numbers for various ξ (calculated from the solutions of PDECOL using π_2). Also, from this table, one can see that the results compare well with those published elsewhere [30].

TABLE 4.7 Further Results of Eq. (4.72)

ξ	Sh*	Sh (PDECOL)
5(−5)	80.75	80.73
1(−4)	57.39	57.38
3(−4)	33.56	33.55
5(−4)	26.22	26.22
8(−4)	20.95	20.94
1(−3)	18.85	18.84

* From Chapter 7 of [30].

PROBLEMS

1. A diffusion-convection problem can be described by the following PDE:

$$\frac{\partial \theta}{\partial t} = \frac{\partial^2 \theta}{\partial x^2} - \lambda \frac{\partial \theta}{\partial x}, \qquad 0 < x < 1, \qquad t > 0$$

with

$$\theta(0, t) = 1, \qquad t > 0$$

$$\frac{\partial \theta}{\partial x}(1, t) = 0, \qquad t > 0$$

$$\theta(x, 0) = 0, \qquad 0 < x < 1$$

where λ = constant.

(a) Discretize the space variable, x, using Galerkin's method with the Hermite cubic basis and set up the MOL-IVP.

(b) Solve the IVP generated in (a) with $\lambda = 25$.

Physically, the problem can represent fluid, heat, or mass flow in which an initially discontinuous profile is propagated by diffusion and convection, the latter with a speed of λ.

2. Given the diffusion-convection equation in Problem 1:

(a) Discretize the space variable, x, using collocation with the Hermite cubic basis and set up the MOL-IVP.

(b) Solve the IVP generated in (a) with $\lambda = 25$.

3. Consider the mass transfer of substance A between a gas phase into solvent I and then into solvent II (the two solvents are assumed to be entirely immiscible). A schematic of the system is shown in Figure 4.10. It is

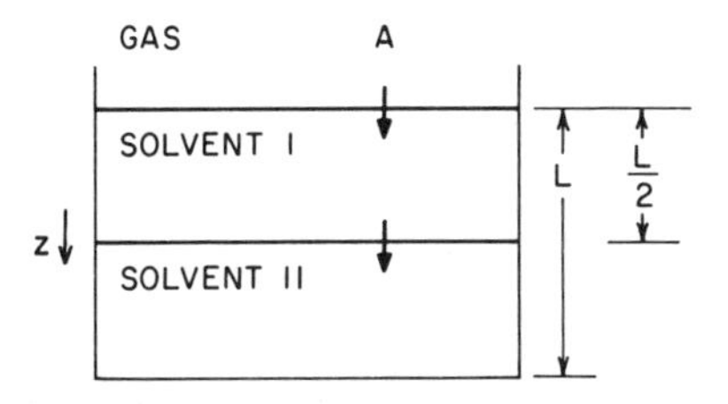

FIGURE 4.10 Immiscible solvent system.

assumed that the concentration of A is sufficiently small so that Fick's law can be used to describe the diffusion in the solvents. The governing differential equations are

$$\frac{\partial C_A^{I}}{\partial t} = \mathscr{D}_{I} \frac{\partial^2 C_A^{I}}{\partial z^2} \qquad \text{(solvent I)}$$

$$\frac{\partial C_A^{II}}{\partial t} = \mathscr{D}_{II} \frac{\partial^2 C_A^{II}}{\partial z^2} \qquad \text{(solvent II)}$$

where

$$C_A^i = \text{concentration of A in region } i \text{ (moles/cm}^3\text{)}$$
$$\mathscr{D}_i = \text{diffusion coefficient in region } i \text{ (cm}^2\text{/s)}$$
$$t = \text{time (s)}$$

with

$$C_A^{I} = C_A^{II} = 0 \quad \text{at} \quad t = 0$$

$$\frac{\partial C_A^{II}}{\partial z} = 0 \quad \text{at} \quad z = L$$

$$-\mathscr{D}_I \frac{\partial C^I}{\partial z} = -\mathscr{D}_{II} \frac{\partial C_A^{II}}{\partial z^2} \quad \text{at} \quad z = \frac{L}{2}$$

$$C_A^{I} = C_A^{II} \quad \text{at} \quad z = L/2 \qquad \text{(distribution coefficient = 1)}$$

$$P_A = HC_A^{I} \quad \text{at} \quad z = 0 \left(\begin{array}{l} P_A = \text{partial pressure of A in} \\ \text{the gas phase; } H \text{ is a constant} \end{array}\right)$$

Compute the concentration profile of A in the liquid phase from $t = 0$ to steady state when $P_A = 1$ atm and $H = 10^4$ atm/(moles·cm^3), for $D_I/D_{II} = 1.0$.

4. Compute the concentration profile of A in the liquid phase (from Problem 3) using $D_I/D_{II} = 10.0$ and $D_I/D_{II} = 0.1$. Compare these profiles to the case where $D_I/D_{II} = 1.0$.

5*. Diffusion and adsorption in an idealized pore can be represented by the PDEs:

$$\frac{\partial c}{\partial t} = D\frac{\partial^2 c}{\partial x^2} - [k_a(1 - f)c - k_d f], \qquad 0 < x < 1, \qquad t > 0$$

$$\frac{\partial f}{\partial t} = \beta[k_a(1 - f)c - k_d f]$$

with

$$c(0, t) = 1, \qquad t > 0$$

$$\frac{\partial c}{\partial x}(1, t) = 0, \qquad t > 0$$

$$c(x, 0) = f(x, 0) = 0 \qquad 0 < x < 1$$

where

c = dimensionless concentration of adsorbate in the fluid within the pore
f = fraction of the pore covered by the adsorbate
x = dimensionless spatial coordinate
k_a = rate constant for adsorption
k_d = rate constant for desorption
D, β = constants

Let $D = \beta = k_a = 1$ and solve the above problem with $k_a/k_d = 0.1, 1.0,$ and 10.0. From these results, discuss the effect of the relative rate of adsorption versus desorption.

6*. The time-dependent material and energy balances for a spherical catalyst pellet can be written as (notation is the same as was used in the section Finite Differences—High-Order Time Approximations):

$$\frac{\partial y}{\partial \tau} = \frac{1}{x^2}\frac{\partial}{\partial x}\left(x^2 \frac{\partial y}{\partial x}\right) - \phi^2 \mathscr{R}$$

$$\text{Le}\frac{\partial \theta}{\partial \tau} = \frac{1}{x^2}\frac{\partial}{\partial x}\left(x^2 \frac{\partial \theta}{\partial x}\right) + \beta\phi^2 \mathscr{R}$$

with

$$\frac{\partial y}{\partial x} = \frac{\partial \theta}{\partial x} = 0 \quad \text{at} \quad x = 0$$

$$y = \theta = 1 \quad \text{at} \quad x = 1$$

$$y = 0, \theta = 1 \quad \text{at} \quad \tau = 0, \qquad \text{for all } x$$

where

$$\mathscr{R} = \exp\left[\gamma\left(1 - \frac{1}{\theta}\right)\right]y \qquad \text{(first order)}$$

Let $\phi = 1.0$, $\beta = 0.04$, $\gamma = 18$, and solve this problem using:

(a) Le = 10
(b) Le = 1
(c) Le = 0.1

Discuss the physical significance of each solution.

7*. Froment [31] developed a two-dimensional model for a fixed-bed reactor where the highly exothermic reactions

$$\begin{array}{ccccc} o\text{-xylene} & \xrightarrow{k_1} & \text{phthalic anhydride} & \xrightarrow{k_2} & CO_2,\ CO,\ H_2O \\ (A) & & (B) & & (C) \end{array}$$

$$(A) \xrightarrow{k_3} \underset{(C)}{CO_2,\ CO,\ H_2O}$$

were occurring on a vanadium catalyst. The steady-state material and energy balances are:

$$\frac{\partial x_1}{\partial z} = \text{Pe}\left[\frac{\partial^2 x_1}{\partial r^2} + \frac{1}{r}\frac{\partial x_1}{\partial r}\right] + \beta_1 R_1, \qquad 0 < r < 1, \qquad 0 < z < 1$$

$$\frac{\partial x_2}{\partial z} = \text{Pe}\left[\frac{\partial^2 x_2}{\partial r^2} + \frac{1}{r}\frac{\partial x_2}{\partial r}\right] + \beta_1 R_2$$

$$\frac{\partial \theta}{\partial z} = \text{Bo}\left[\frac{\partial^2 \theta}{\partial r^2} + \frac{1}{r}\frac{\partial \theta}{\partial r}\right] + \beta_2 R_1 + \beta_3 R_2$$

with

$$x_1 = x_2 = 0 \quad \text{and} \quad \theta = \theta_0 \quad \text{at} \quad z = 0, \qquad 0 < r < 1$$

$$\frac{\partial x_1}{\partial r} = \frac{\partial x_2}{\partial r} = \frac{\partial \theta}{\partial r} = 0 \quad \text{at} \quad r = 0, \qquad 0 < z < 1$$

$$\frac{\partial x_1}{\partial r} = \frac{\partial x_2}{\partial r} = 0 \quad \text{and} \quad \frac{\partial \theta}{\partial r} = \text{Bi}\,(\theta - \theta_w) \quad \text{at} \quad r = 1, \qquad 0 < z < 1$$

where

x_1 = fractional conversion to B
x_2 = fractional conversion to C
θ = dimensionless temperature
z = dimensionless axial coordinate
r = dimensionless radial coordinate
$R_1 = k_1(1 - x_1 - x_2) - k_2 x_1$
$R_2 = k_2 x_1 + k_3(1 - x_1 - x_2)$
Pe, Bo = constants
β_i = constants, $i = 1, 2, 3$
Bi = Biot number
θ_w = dimensionless wall temperature

Froment gives the data required to specify this system of PDEs, and one set of his data is given in Chapter 8 of [30] as:

$$\mathrm{Pe} = 5.76, \qquad \mathrm{Bo} = 10.97, \qquad \mathrm{Bi} = 2.5$$

$$\beta_1 = 5.106, \qquad \beta_2 = 3.144, \qquad \beta_3 = 11.16$$

where

$$k_i = \exp\left[a_i + \gamma_i\left(1 - \frac{1}{\theta}\right)\right], \qquad i = 1, 2, 3$$

with

$$a_1 = -1.74, \qquad a_2 = -4.24, \qquad a_3 = -3.89$$

$$\gamma_1 = 21.6, \qquad \gamma_2 = 25.1, \qquad \gamma_3 = 22.9$$

Let $\theta_w = \theta_0 = 1$ and solve the reactor equations with:

(a) $a_1 = -1.74$

(b) $a_1 = -0.87$

(c) $a_1 = -3.48$

Comment on the physical implications of your results.

8*. The simulation of transport with chemical reaction in the stratosphere presents interesting problems, since certain reactions are photochemical, i.e., they require sunlight to proceed. In the following problem let C_1 denote the concentration of ozone, O_3, C_2 denote the concentration of oxygen singlet, O, and C_3 denote the concentration of oxygen, O_2 (assumed to be constant). If z specifies the altitude in kilometers, and a Fickian model of turbulent eddy diffusion (neglecting convection) is used to describe the transport of chemical species, then the continuity equations of the given species are

$$\frac{\partial C_1}{\partial t} = \frac{\partial}{\partial z}\left[K\frac{\partial C_1}{\partial z}\right] + R_1, \qquad 30 < z < 50, \quad 0 < t$$

$$\frac{\partial C_2}{\partial t} = \frac{\partial}{\partial z}\left[K\frac{\partial C_2}{\partial z}\right] + R_2$$

with

$$\frac{\partial C_1}{\partial z}(30, t) = \frac{\partial C_2}{\partial z}(30, t) = 0, \qquad t > 0$$

$$\frac{\partial C_1}{\partial z}(50, t) = \frac{\partial C_2}{\partial z}(50, t) = 0, \qquad t > 0$$

$$C_1(z, 0) = 10^6\gamma(z), \qquad 30 \leq z \leq 50$$

$$C_2(z, 0) = 10^{12}\gamma(z), \qquad 30 \leq z \leq 50$$

where

$$
\begin{aligned}
K &= \exp\left[\frac{z}{5}\right] \\
R_1 &= -k_1C_1C_3 - k_2C_1C_2 + 2k_3(t)C_3 + k_4(t)C_2 \\
R_2 &= k_1C_1C_3 - k_2C_1C_2 - k_4(t)C_2 \\
C_3 &= 3.7 \times 10^{16} \\
k_1 &= 1.63 \times 10^{-16} \\
k_2 &= 4.66 \times 10^{-16}
\end{aligned}
$$

$$
k_i(t) = \begin{cases} exp\left[\dfrac{-\nu_i}{sin\ (\omega t)}\right], & \text{for } \sin(\omega t) > 0, \\ 0, & \text{for } \sin(\omega t) \leq 0, \end{cases} \qquad i = 3, 4
$$

$$
\begin{aligned}
\nu_3 &= 22.62 \\
\nu_4 &= 7.601 \\
\omega &= \frac{\pi}{43{,}200} \\
\gamma(z) &= 1 - \left[\frac{z - 40}{10}\right]^2 + \frac{1}{2}\left[\frac{z - 40}{10}\right]^4
\end{aligned}
$$

Notice that the reaction constants $k_3(\mathrm{t})$ and $k_4(t)$ build up to a peak at noon ($t = 21{,}600$) and are switched off from sunset ($t = 43{,}200$) to sunrise ($t = 86{,}400$). Thus, these reactions model the diurnal effect. Calculate the concentration of ozone for a 24-h time period, and compare your results with those given in [26].

REFERENCES

1. Liskovets, O. A., "The Method of Lines," J. Diff. Eqs., *1,* 1308 (1965).
2. Mitchell, A. R., and D. F. Griffiths, *The Finite Difference Method in Partial Differential Equations,* Wiley, Chichester (1980).
3. Wilkes, J. O., "In Defense of the Crank-Nicolson Method," A.I.Ch.E. J., *16,* 501 (1970).
4. Douglas, J., Jr., and B. F. Jones, "On Predictor-Corrector Method for Nonlinear Parabolic Differential Equations," J. Soc. Ind. Appl. Math., *11,* 195 (1963).

5. Lees, M., "An Extrapolated Crank-Nicolson Difference Scheme for Quasi-Linear Parabolic Equations," in *Nonlinear Partial Differential Equations,* W. F. Ames (ed.), Academic, New York (1967).
6. Davis, M. E., "Analysis of an Annular Bed Reactor for the Methanation of Carbon Oxides," Ph.D. Thesis, Univ. of Kentucky, Lexington (1981).
7. Davis, M. E., G. Fairweather and J. Yamanis, "Annular Bed Reactor-Methanation of Carbon Dioxide," Can. J. Chem. Eng., *59,* 497 (1981).
8. Sepehrnoori, K., and G. F. Carey, "Numerical Integration of Semidiscrete Evolution Systems," Comput. Methods Appl. Mech. Eng., *27,* 45 (1981).
9. Shampine, L. F., and M. K. Gordon, *Computer Solution of Ordinary Differential Equations: The Initial Value Problem,* Freeman, San Francisco (1975).
10. International Mathematics and Statistics Libraries Inc., Sixth Floor—NBC Building, 7500 Bellaire Boulevard, Houston, Tex.
11. Verwer, J. G., "Algorithm 553-M3RK, An Explicit Time Integrator for Semidiscrete Parabolic Equations," ACM TOMS, *6,* 236 (1980).
12. Skeel, R. D., and A. K. Kong, "Blended Linear Multistep Methods," ACM TOMS. Math. Software, *3,* 326 (1977).
13. Kehoe, J. P. C., and J. B. Butt, "Kinetics of Benzene Hydrogenation by Supported Nickel at Low Temperature," J. Appl. Chem. Biotechnol., *22,* 23 (1972).
14. Hindmarsh, A. C., "LSODE and LSODI, Two New Initial Value Ordinary Differential Equation Solvers," ACM SIGNUM Newsletter, December (1980).
15. Kehoe, J. P. C., and J. B. Butt, "Transient Response and Stability of a Diffusionally Limited Exothermic Catalytic Reaction," 5th International Reaction Engineering Symposium, Amsterdam (1972).
16. Hindmarsh, A. C., "GEARIB: Solution of Implicit Systems of Ordinary Differential Equations with Banded Jacobian," Lawrence Livermore Laboratory Report UCID-30103 (1976).
17. Hindmarsh, A. C., "GEAR: Ordinary Differential Equation System Solver," Lawrence Livermore Laboratory Report UCID-30001 (1974).
18. Hindmarsh, A. C., "GEARB: Solution of Ordinary Differential Equations Having Banded Jacobians," Lawrence Livermore Laboratory Report UCID-30059 (1975).
19. Dew, P. M., and Walsh, J. E., "A Set of Library Routines for Solving Parabolic Equations in One Space Variables," ACM TOMS, *7,* 295 (1981).
20. Harwell Subroutine Libraries, Computer Science and Systems Division of the United Kingdom Atomic Energy Authority, Harwell, England.
21. Sewell, G., "IMSL Software for Differential Equations in One Space Variable," IMSL Tech. Report No. 8202 (1982).

22. Sincovec, R. F., and N. K. Madsen, "PDEPACK: Partial Differential Equations Package Users Guide," Scientific Computing Consulting Services, Colorado Springs, Colo. (1981).

23. Schiesser, W., "DDS/2—An Introduction to the Numerical Method of Lines Integration of Partial Differential Equations," Lehigh Univ., Bethlehem, Pa. (1976).

24. Hyman, J. M., "MOL1D: A General-Purpose Subroutine Package for the Numerical Solution of Partial Differential Equations," Los Alamos Scientific Laboratory Report LA-7595-M, March (1979).

25. Madsen, N. K., and R. F. Sincovec, "PDECOL, General Collocation Software for Partial Differential Equations," ACM TOMS, *5,* 326 (1979).

26. Leaf, G. K., M. Minkoff, G. D. Byrne, D. Sorensen, T. Blecknew, and J. Saltzman, "DISP: A Software Package for One and Two Spatially Dimensioned Kinetics-Diffusion Problems," Report ANL-77-12, Argonne National Laboratory, Argonne, Ill. (1977).

27. Schryer, N. L., "Numerical Solution of Time-Varying Partial Differential Equations in One Space Variable," Computer Science Tech. Report 53, Bell Laboratories, Murray Hill, N.J. (1977).

28. Fox, P. A., A. D. Hall, and N. L. Schryer, "The PORT Mathematical Subroutine Library," ACM TOMS, *4,* 104 (1978).

29. Carver, M., "The FORSIM VI Simulation Package for the Automatic Solution of Arbitrarily Defined Partial Differential and/or Ordinary Differential Equation Systems," Rep. AECL-5821, Chalk River Nuclear Laboratories, Ontario, Canada (1978).

30. Villadsen, J., and M. L. Michelsen, *Solution of Differential Equation Models by Polynomial Approximation,* Prentice-Hall, Englewood Cliffs, N.J. (1978).

31. Froment, G. F., "Fixed Bed Catalytic Reactors," Ind. Eng. Chem., *59,* 18 (1967).

BIBLIOGRAPHY

An overview of finite diffrence and finite element methods for parabolic partial differential equations in time and one space dimension has been given in this chapter. For additional or more detailed information, see the following texts:

Finite Difference

Ames, W. F., *Nonlinear Partial Differential Equations in Engineering,* Academic, New York (1965).

Ames, W. F., ed., *Nonlinear Partial Differential Equations,* Academic, New York (1967).

Ames, W. F., *Numerical Methods for Partial Differential Equations,* 2nd ed., Academic, New York (1977).

Finlayson, B. A., *Nonlinear Analysis in Chemical Engineering,* McGraw-Hill, New York (1980).

Forsythe, G. E., and W. R. Wason, *Finite Difference Methods for Partial Differential Equations,* Wiley, New York (1960).

Issacson, I., and H. B. Keller, *Analysis of Numerical Methods,* Wiley, New York (1966).

Mitchell, A. R., and D. F. Griffiths, *The Finite Difference Method in Partial Differential Equations,* Wiley, Chichester (1980).

Finite Element

Becker, E. B., G. F. Carey, and J. T. Oden, *Finite Elements: An Introduction,* Prentice-Hall, Englewood Cliffs, N.J. (1981).

Fairweather, G., *Finite Element Galerkin Methods for Differential Equations,* Marcel Dekker, New York (1978).

Mitchell, A. R., and D. F. Griffiths, *The Finite Difference Method in Partial Differential Equations,* Wiley, Chichester (1980). Chapter 5 discusses the Galerkin method.

Mitchell, A. R., and R. Wait, *The Finite Element Method in Partial Differential Equations,* Wiley, New York (1977).

Strang, G., and G. J. Fix, *An Analysis of the Finite Element Method,* Prentice-Hall, Englewood Cliffs, N.J. (1973).

5

Partial Differential Equations in Two Space Variables

INTRODUCTION

In Chapter 4 we discussed the various classifications of PDEs and described finite difference (FD) and finite element (FE) methods for solving parabolic PDEs in one space variable. This chapter begins by outlining the solution of elliptic PDEs using FD and FE methods. Next, parabolic PDEs in two space variables are treated. The chapter is then concluded with a section on mathematical software, which includes two examples.

ELLIPTIC PDES—FINITE DIFFERENCES

Background

Let R be a bounded region in the $x - y$ plane with boundary ∂R. The equation

$$\frac{\partial}{\partial x}\left[a_1(x, y)\,\frac{\partial w}{\partial x}\right] + \frac{\partial}{\partial y}\left[a_2(x, y)\,\frac{\partial w}{\partial y}\right] = d\left(x, y, w, \frac{\partial w}{\partial x}, \frac{\partial w}{\partial y}\right)$$

$$a_1 a_2 > 0 \tag{5.1}$$

is elliptic in R (see Chapter 4 for the definition of elliptic equations), and three problems involving (5.1) arise depending upon the subsidiary conditions prescribed on ∂R:

1. Dirichlet problem:

$$w = f(x, y) \quad \text{on} \quad \partial R \tag{5.2}$$

2. Neumann problem:

$$\frac{\partial w}{\partial n} = g(x, y) \quad \text{on} \quad \partial R \tag{5.3}$$

where $\partial/\partial n$ refers to differentiation along the outward normal to ∂R

3. Robin problem:

$$\alpha(x, y)w + \beta(x, y)\frac{\partial w}{\partial n} = \gamma(x, y) \quad \text{on} \quad \partial R \tag{5.4}$$

We illustrate these three problems on Laplace's equation in a square.

Laplace's Equation in a Square

Laplace's equation is

$$\frac{\partial^2 w}{\partial x^2} + \frac{\partial^2 w}{\partial y^2} = 0, \qquad 0 \leqslant x \leqslant 1, \quad 0 \leqslant y \leqslant 1 \tag{5.5}$$

Let the square region R, $0 \leqslant x \leqslant 1$, $0 \leqslant y \leqslant 1$, be covered by a grid with sides parallel to the coordinate axis and grid spacings such that $\Delta x = \Delta y = h$. If $Nh = 1$, then the number of internal grid points is $(N - 1)^2$. A second-order finite difference discretization of (5.5) at any interior node is:

$$\frac{1}{(\Delta x)^2}[u_{i+1,j} - 2u_{i,j} + u_{i-1,j}] + \frac{1}{(\Delta y)^2}[u_{i,j+1} - 2u_{i,j} + u_{i,j-1}] = 0 \tag{5.6}$$

where

$$\begin{aligned} u_{i,j} &\simeq w(x_i, y_j) \\ x_i &= ih \\ y_j &= jh \end{aligned}$$

Since $\Delta x = \Delta y$, (5.6) can be written as:

$$u_{i,j-1} + u_{i+1,j} - 4u_{i,j} + u_{i-1,j} + u_{i,j+1} = 0 \tag{5.7}$$

with an error of $0(h^2)$.

Dirichlet Problem If $w = f(x, y)$ on ∂R, then

$$u_{i,j} = f(x_i, y_j) \tag{5.8}$$

for (x_i, y_j) on ∂R. Equations (5.7) and (5.8) completely specify the discretization, and the ensuing matrix problem is

$$\mathbf{Au} = \mathbf{f} \tag{5.9}$$

where

$$
A = \begin{bmatrix} J & -I & & & & 0 \\ -I & \cdot & \cdot & & & \\ & \cdot & \cdot & \cdot & & \\ & & \cdot & \cdot & \cdot & \\ & & & \cdot & \cdot & -I \\ 0 & & & & -I & J \end{bmatrix}, \qquad (N-1)^2 \times (N-1)^2
$$

$$
I = \text{identity matrix}, \qquad (N-1) \times (N-1)
$$

$$
J = \begin{bmatrix} 4 & -1 & & & 0 \\ -1 & \cdot & \cdot & & \\ & \cdot & \cdot & \cdot & \\ & & \cdot & \cdot & -1 \\ 0 & & & -1 & 4 \end{bmatrix}, \qquad (N-1) \times (N-1)
$$

$$
\mathbf{u} = [u_{1,1}, \ldots, u_{N-1,1}, u_{1,2}, \ldots, u_{N-1,2}, \ldots, u_{1,N-1}, \ldots, u_{N-1,N-1}]^T
$$

$$
\begin{aligned}
\mathbf{f} = [&f(0, y_1) + f(x_1, 0), f(x_2, 0), \ldots, f(x_{N-1}, 0) \\
&+ f(1, y_1), f(0, y_2), 0, \ldots, 0, f(1, y_2), \ldots, f(0, y_{N-1}) \\
&+ f(x_1, 1), f(x_1, 1), f(x_2, 1), \ldots, f(x_{N-1}, 1) + f(1, y_{N-1})]^T
\end{aligned}
$$

Notice that the matrix A is block tridiagonal and that most of its elements are zero. Therefore, when solving problems of this type, a matrix-solving technique that takes into account the sparseness and the structure of the matrix should be used. A few of these techniques are outlined in Appendix E.

Neumann Problem Discretize (5.3) using the method of false boundaries to give:

$$
\frac{1}{2h}[u_{-1,j} - u_{1,j}] = g_{0,j}
$$

or **(5.10)**

$$
\frac{1}{2h}[u_{i,-1} - u_{i,1}] = g_{i,0}
$$

where

$$
g_{0,j} = g(0, jh)
$$

Combine (5.10) and (5.7) with the result

$$Au = 2h\mathbf{g} \tag{5.11}$$

where

$$A = \begin{bmatrix} K & -2I & & & \\ -I & \cdot & \cdot & & \\ & \cdot & \cdot & \cdot & \\ & & \cdot & \cdot & \cdot \\ & & & \cdot & \cdot & -I \\ & & & & -2I & K \end{bmatrix}, \qquad (N+1)^2 \times (N+1)^2$$

$$K = \begin{bmatrix} 4 & -2 & & & \\ -1 & 4 & -1 & & \\ -1 & \cdot & \cdot & & \\ & \cdot & \cdot & \cdot & \\ & & \cdot & \cdot & \cdot \\ & & \cdot & -1 & 4 & -1 \\ & & & & -2 & 4 \end{bmatrix}, \qquad (N+1) \times (N+1)$$

I = identity matrix, $(N+1) \times (N+1)$

$$\mathbf{u} = [u_{0,0}, \ldots, u_{N,0}, u_{0,1}, \ldots, u_{N,1}, \ldots, u_{0,N}, \ldots, u_{N,N}]^T$$

$$\mathbf{g} = [2g_{0,0}, g_{1,0}, \ldots, 2g_{N,0}, g_{0,1}, 0, \ldots, 0, g_{N,1}, \ldots,$$
$$2g_{0,N}, g_{1,N}, \ldots, g_{N-1,N}, 2g_{N,N}]^T$$

In contrast to the Dirichlet problem, the matrix A is now singular. Thus A has only $(N+1)^2 - 1$ rows or columns that are linearly independent. The solution of (5.11) therefore involves an arbitrary constant. This is a characteristic of the solution of a Neumann problem.

Robin Problem. Consider the boundary conditions of form

$$\left.\begin{aligned} \frac{\partial w}{\partial x} - \phi_1 w &= f_0(y), \quad x = 0 \\ \frac{\partial w}{\partial x} + \eta_1 w &= f_1(y), \quad x = 1 \end{aligned}\right\} \quad \text{for } 0 \leq y \leq 1$$

$$\left.\begin{aligned} \frac{\partial w}{\partial y} - \phi_2 w &= g_0(x), \quad y = 0 \\ \frac{\partial w}{\partial y} + \eta_2 w &= g_1(x), \quad y = 1 \end{aligned}\right\} \quad \text{for } 0 \leq x \leq 1 \tag{5.12}$$

where ϕ and η are constants and f and g are known functions. Equations (5.12) can be discretized, say by the method of false boundaries, and then included in the discretization of (5.5). During these discretizations, it is important to maintain the same order of accuracy in the boundary discretization as with the PDE discretization. The resulting matrix problem will be $(N + 1)^2 \times (N + 1)^2$, and its form will depend upon (5.12).

Usually, a practical problem contains a combination of the different types of boundary conditions, and their incorporation into the discretization of the PDE can be performed as stated above for the three cases.

EXAMPLE 1

Consider a square plate $R = \{(x, y): 0 \leqslant x \leqslant 1, 0 \leqslant y \leqslant 1\}$ with the heat conduction equation

$$\frac{\partial^2 T}{\partial x^2} + \frac{\partial^2 T}{\partial y^2} = 0$$

Set up the finite difference matrix problem for this equation with the following boundary conditions:

$$
\begin{aligned}
T(x, y) = T(0, y) &= T_1 && \text{(fixed temperature)} \\
T(1, y) &= T_2 && \text{(fixed temperature)} \\
\frac{\partial T}{\partial y}(x, 0) &= 0 && \text{(insulated surface)} \\
\frac{\partial T}{\partial y}(x, 1) &= k[T(x, 1) - T_2] && \text{(heat convected away at } y = 1)
\end{aligned}
$$

where T_1, T_2, and k are constants and $T_1 \geqslant T(x, y) \geqslant T_2$.

SOLUTION

Impose a grid on the square region R such that $x_i = ih$, $y_j = jh$ $(\Delta x = \Delta y)$ and $Nh = 1$. For any interior grid point

$$u_{i,j-1} + u_{i+1,j} - 4u_{i,j} + u_{i-1,j} + u_{i,j+1} = 0$$

where

$$u_{i,j} \simeq T(x_i, y_j)$$

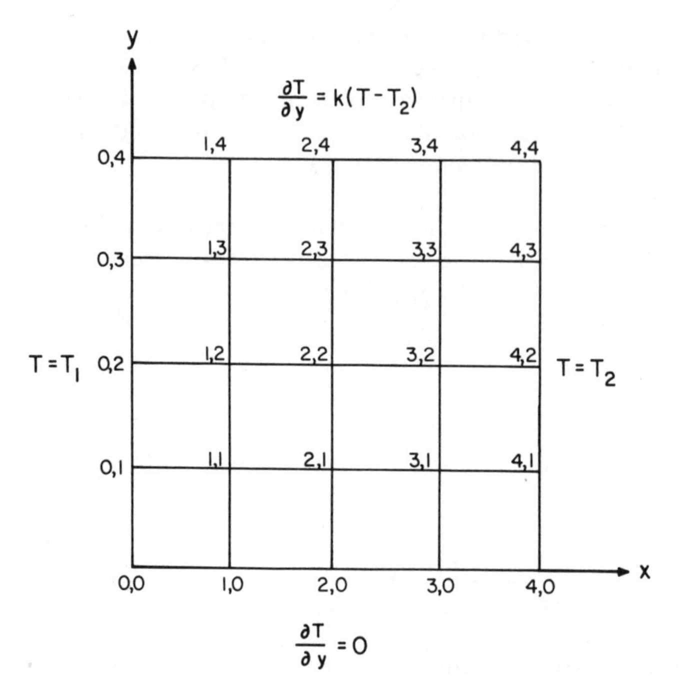

FIGURE 5.1 Grid for Example 1.

At the boundaries $x = 0$ and $x = 1$ the boundary conditions are Dirichlet. Therefore

$$u_{0,j} = T_1, \qquad \text{for } j = 0, \ldots, N$$

$$u_{N,j} = T_2, \qquad \text{for } j = 0, \ldots, N$$

At $y = 0$ the insulated surface gives rise to a Neumann condition that can be discretized as

$$u_{i,-1} - u_{i,1} = 0, \qquad \text{for } i = 1, \ldots, N - 1$$

and at $y = 1$ the Robin condition is

$$\frac{u_{i,N-1} - u_{i,N+1}}{2h} = k[u_{i,N} - T_2], \qquad \text{for } i = 1, \ldots, N - 1$$

If $N = 4$, then the grid is as shown in Figure 5.1 and the resulting matrix problem is:

$$
\begin{bmatrix}
-4 & 1 & & 2 & & & & & & & & & & & \\
1 & -4 & 1 & & 2 & & & & & & & & & & \\
& 1 & -4 & & & 2 & & & & & & & & & \\
1 & & & -4 & 1 & & 1 & & & & & & & & \\
& 1 & & 1 & -4 & 1 & & 1 & & & & & & & \\
& & 1 & & 1 & -4 & & & 1 & & & & & & \\
& & & 1 & & & -4 & 1 & & 1 & & & & & \\
& & & & 1 & & 1 & -4 & 1 & & 1 & & & & \\
& & & & & 1 & & 1 & -4 & & & 1 & & & \\
& & & & & & 1 & & & -4 & 1 & & 1 & & \\
& & & & & & & 1 & & 1 & -4 & 1 & & 1 & \\
& & & & & & & & 1 & & 1 & -4 & & & 1 \\
& & & & & & & & & 2 & & & (4+2hk) & 1 & \\
& & & & & & & & & & 2 & & 1 & -(4+2hk) & 1 \\
& & & & & & & & & & & 2 & & 1 & -(4+2hk)
\end{bmatrix}
\begin{bmatrix}
u_{1,0} \\ u_{2,0} \\ u_{3,0} \\ u_{1,1} \\ u_{2,1} \\ u_{3,1} \\ u_{1,2} \\ u_{2,2} \\ u_{3,2} \\ u_{1,3} \\ u_{2,3} \\ u_{3,3} \\ u_{1,4} \\ u_{2,4} \\ u_{3,4}
\end{bmatrix}
=
\begin{bmatrix}
-T_1 \\ 0 \\ -T_2 \\ -T_1 \\ 0 \\ -T_2 \\ -T_1 \\ 0 \\ -T_2 \\ -T_1 \\ 0 \\ -T_2 \\ -(T_1+2hkT_2) \\ -2hkT_2 \\ -(T_2+2hkT_2)
\end{bmatrix}
$$

Notice how the boundary conditions are incorporated into the matrix problem. The matrix generated by the finite difference discretization is sparse, and an appropriate linear equation solver should be employed to determine the solution. Since the error is $0(h^2)$, the error in the solution with $N = 4$ is $0(0.0625)$. To obtain a smaller error, one must increase the value of N, which in turn increases the size of the matrix problem.

Variable Coefficients and Nonlinear Problems

Consider the following elliptic PDE:

$$-(P(x, y)w_x)_x - (P(x, y)w_y)_y + \eta(x, y)w^\sigma = f(x, y) \tag{5.13}$$

defined on a region R with boundary ∂R and

$$a(x, y)w + b(x, y)\frac{\partial w}{\partial n} = c(x, y), \qquad \text{for } (x, y) \text{ on } \partial R \tag{5.14}$$

Assume that P, P_x, P_y, η, and f are continuous in R and

$$\begin{aligned} P(x, y) &> 0 \\ \eta(x, y) &> 0 \end{aligned} \tag{5.15}$$

Also, assume a, b, and c are piecewise continuous and

$$\left.\begin{aligned} a(x, y) &\geq 0 \\ b(x, y) &\geq 0 \\ a + b &> 0 \end{aligned}\right\} \qquad \text{for } (x, y) \text{ on } \partial R \tag{5.16}$$

If $\sigma = 1$, then (5.13) is called a self-adjoint elliptic PDE because of the form of the derivative terms. A finite difference discretization of (5.13) for any interior node is

$$-\delta_x(P(x_i, y_j)\delta_x u_{i,j}) - \delta_y(P(x_i, y_j)\delta_y u_{i,j}) + \eta(x_i, y_j)u_{i,j}^\sigma = f(x_i, y_j) \tag{5.17}$$

where

$$\delta_x u_{i,j} = \frac{u_{i+1/2,j} - u_{i-1/2,j}}{\Delta x}$$

$$\delta_y u_{i,j} = \frac{u_{i,j+1/2} - u_{i,j-1/2}}{\Delta y}$$

The resulting matrix problem will still remain in block-tridiagonal form, but if $\sigma \neq 0$ or 1, then the system is nonlinear. Therefore, a Newton iteration must be performed. Since the matrix problem is of considerable magnitude, one would like to minimize the number of Newton iterations to obtain solution. This is the rationale behind the Newton-like methods of Bank and Rose [1]. Their methods try to accelerate the convergence of the Newton method so as to minimize the amount of computational effort in obtaining solutions from large systems of nonlinear algebraic equations. A problem of practical interest, the simulation of a two-phase, cross-flow reactor (three nonlinear coupled elliptic PDEs), was solved in [2] using the methods of Bank and Rose, and it was shown that these methods significantly reduced the number of iterations required for solution [3].

Nonuniform Grids

Up to this point we have limited our discussions to uniform grids, i.e., $\Delta x = \Delta y$. Now let $k_j = y_{j+1} - y_j$ and $h_i = x_{i+1} - x_i$. Following the arguments of Varga [4], at each interior mesh point (x_i, y_j) for which $u_{i,j} \simeq w(x_i, y_j)$, integrate (5.17) over a corresponding mesh region $\bar{r}_{i,j}$ ($\sigma = 1$):

$$-\iint_{\bar{r}_{i,j}} \left\{(Pw_x)_x + (Pw_y)_y\right\} dx\, dy + \iint_{\bar{r}_{i,j}} \eta w\, dx\, dy = \iint_{\bar{r}_{i,j}} f\, dx\, dy \qquad \textbf{(5.18)}$$

By Green's theorem, any two differentiable functions $s(x, y)$ and $t(x, y)$ defined in $\bar{r}_{i,j}$ obey

$$\iint_{\bar{r}_{i,j}} (s_x - t_y)\, dx\, dy = \int_{\partial\bar{r}_{i,j}} (t\, dx + s\, dy) \qquad \textbf{(5.19)}$$

where $\partial\bar{r}_{i,j}$ is the boundary of $\bar{r}_{i,j}$ (refer to Figure 5.2). Therefore, (5.18) can be written as

$$-\int_{\partial\bar{r}_{i,j}} \left\{Pw_x\, dy - Pw_y\, dx\right\} + \iint_{\bar{r}_{i,j}} \eta w\, dx\, dy = \iint_{\bar{r}_{i,j}} f\, dx\, dy \qquad \textbf{(5.20)}$$

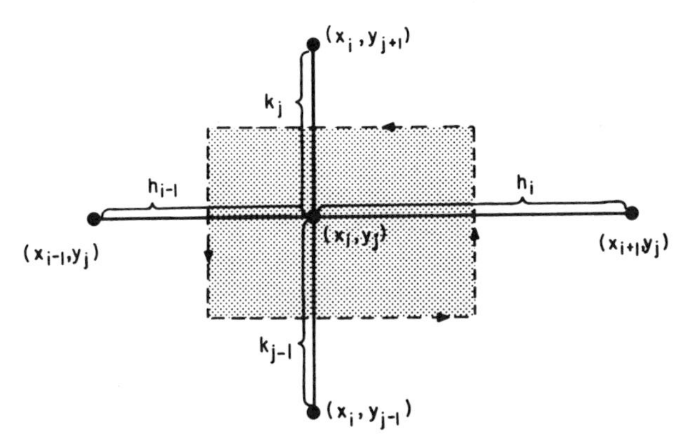

FIGURE 5.2 Nonuniform grid spacing (shaded area is the integration area). Adapted from Richard S. Varga, *Matrix Iterative Analysis,* copyright © 1962, p. 184. Reprinted by permission of Prentice-Hall, Inc., Englewood Cliffs, N. J.

The double integrals above can be approximated by

$$\iint_{\bar{r}_{i,j}} z\,dx\,dy \simeq A_{i,j}\, z_{i,j} \tag{5.21}$$

for any function $z(x, y)$ such that $z(x_i, y_j) = z_{i,j}$ and

$$A_{i,j} = \frac{(h_{i-1} + h_i)(k_{j-1} + k_j)}{4}$$

The line integral in (5.20) is approximated by central differences (integration follows arrows on Figure 5.2). For example, consider the portion of the line integral from $(x_{i+1/2}, y_{j-1/2})$ to $(x_{i+1/2}, y_{j+1/2})$:

$$-\int_{(x_{i+\frac{1}{2}},y_{j-\frac{1}{2}})}^{(x_{i+\frac{1}{2}},y_{j+\frac{1}{2}})} (Pw_x dy - Pw_y dx)$$

$$= \left(\frac{k_{j-1}}{2} P_{i+\frac{1}{2},j-\frac{1}{2}} + \frac{k_j}{2} P_{i+\frac{1}{2},j+\frac{1}{2}}\right)\left(\frac{u_{i,j} - u_{i+1,j}}{h_i}\right) \tag{5.22}$$

where

$$P_{i+\frac{1}{2},j-\frac{1}{2}} = P(x_{i+\frac{1}{2}}, y_{j-\frac{1}{2}})$$

Therefore, the complete line integral is approximated by

$$\begin{aligned}
&\left(\frac{k_{j-1}}{2} P_{i+\frac{1}{2},j-\frac{1}{2}} + \frac{k_j}{2} P_{i+\frac{1}{2},j+\frac{1}{2}}\right)\left(\frac{u_{i,j} - u_{i+1,j}}{h_i}\right) \\
&\quad + \left(\frac{k_{j-1}}{2} P_{i-\frac{1}{2},j-\frac{1}{2}} + \frac{k_j}{2} P_{i-\frac{1}{2},j+\frac{1}{2}}\right)\left(\frac{u_{i,j} - u_{i-1,j}}{h_{i-1}}\right) \\
&\quad + \left(\frac{h_{i-1}}{2} P_{i-\frac{1}{2},j+\frac{1}{2}} + \frac{h_i}{2} P_{i+\frac{1}{2},j+\frac{1}{2}}\right)\left(\frac{u_{i,j} - u_{i,j+1}}{k_j}\right) \\
&\quad + \left(\frac{h_{i-1}}{2} P_{i-\frac{1}{2},j-\frac{1}{2}} + \frac{h_i}{2} P_{i+\frac{1}{2},j-\frac{1}{2}}\right)\left(\frac{u_{i,j} - u_{i,j-1}}{k_{j-1}}\right)
\end{aligned} \tag{5.23}$$

Using (5.23) and (5.21) on (5.20) gives

$$D_{i,j}u_{i,j} - L_{i,j}u_{i-1,j} - M_{i,j}u_{i+1,j} - T_{i,j}u_{i,j+1} - B_{i,j}u_{i,j-1} = S_{i,j} \tag{5.24}$$

where

$$D_{i,j} = L_{i,j} + M_{i,j} + T_{i,j} + B_{i,j} + \eta_{i,j}\frac{(h_{i-1} + h_i)(k_{j-1} + k_j)}{4}$$

$$h_{i-1}L_{i,j} = \frac{k_{j-1}}{2}P_{i-\frac{1}{2},j-\frac{1}{2}} + \frac{k_j}{2}P_{i-\frac{1}{2},j+\frac{1}{2}}$$

$$h_iM_{i,j} = \frac{k_{j-1}}{2}P_{i+\frac{1}{2},j-\frac{1}{2}} + \frac{k_j}{2}P_{i+\frac{1}{2},j+\frac{1}{2}}$$

$$k_jT_{i,j} = \frac{h_{i-1}}{2}P_{i-\frac{1}{2},j+\frac{1}{2}} + \frac{h_i}{2}P_{i+\frac{1}{2},j+\frac{1}{2}}$$

$$k_{j-1}B_{i,j} = \frac{h_{i-1}}{2}P_{i-\frac{1}{2},j-\frac{1}{2}} + \frac{h_i}{2}P_{i+\frac{1}{2},j-\frac{1}{2}}$$

$$S_{i,j} = f_{i,j}\frac{(h_{i-1} + h_i)(k_{j-1} + k_j)}{4}$$

Notice that if $h_i = h_{i-1} = k_j = k_{j-1}$ and $P(x, y) =$ constant, (5.24) becomes the standard second-order accurate difference formula for (5.17). Also, notice that if $P(x, y)$ is discontinuous at x_i and/or y_j as in the case of inhomogeneous media, (5.24) is still applicable since P is not evaluated at either the horizontal (y_j) or the vertical (x_i) plane. Therefore, the discretization of (5.18) at any interior node is given by (5.24). To complete the discretization of (5.18) requires knowledge of the boundary discretization. This is discussed in the next section.

EXAMPLE 2

In Chapter 4 we discussed the annular bed reactor (see Figure 4.5) with its mass continuity equation given by (4.46). If one now allows for axial dispersion of mass, the mass balance for the annular bed reactor becomes

$$\delta_1\frac{\partial f}{\partial z} = \left[\frac{\text{Am An}}{\text{Re Sc}}\right]\frac{1}{r}\frac{\partial}{\partial r}\left(rD^r\frac{\partial f}{\partial r}\right) + \left[\frac{\text{An/Am}}{\text{Re Sc}}\right]\frac{\partial}{\partial z}\left(D^z\frac{\partial f}{\partial z}\right) + \left[\frac{\text{Am An}}{\text{Re Sc}}\right]\delta_2\phi^2R(f)$$

where the notation is as in (4.46) except for

D^r = dimensionless radial dispersion coefficient
D^z = dimensionless axial dispersion coefficient

At $r = r_{sc}$, the core-screen interface, we assume that the convection term is equal to zero (zero velocity), thus reducing the continuity equation to

$$\frac{1}{r}\frac{\partial}{\partial r}\left(rD^r\frac{\partial f}{\partial r}\right) + \frac{\partial}{\partial z}\left(D^z\frac{\partial f}{\partial z}\right) = 0$$

Also, since the plane $r = r_{sc}$ is an interface between two media, D^r and D^z are discontinuous at this position. Set up the difference equation at the interface $r = r_{sc}$ using the notation of Figure 5.3. If we now consider D^r and D^z to be constants, and let $h_{i-1} = h_i$, and $k_{j-1} = k_j$, show that the interface discretization simplifies to the standard second-order correct discretization.

SOLUTION

Using (5.18) to discretize the PDE at $r = r_{sc}$ gives

$$-\iint_{\bar{r}_{i,j}} \left[\frac{1}{r}\frac{\partial}{\partial r}\left(rD^r \frac{\partial f}{\partial r}\right) + \frac{\partial}{\partial z}\left(D^z \frac{\partial f}{\partial z}\right)\right] r\,dr\,dz = 0$$

Upon applying Green's theorem to this equation, we have

$$-\int_{\partial\bar{r}_{i,j}} \left[rD^r \frac{\partial f}{\partial r}\,dz - D^z \frac{\partial f}{\partial z}\,r\,dr\right] = 0$$

If the line integral is approximated by central differences, then

$$r_{i+\frac{1}{2}}\left(\frac{k_{j-1}}{2}D^r_{i+\frac{1}{2},j-\frac{1}{2}} + \frac{k_j}{2}D^r_{i+\frac{1}{2},j+\frac{1}{2}}\right)\left(\frac{u_{i,j} - u_{i+1,j}}{h_i}\right)$$

$$+\, r_{i-\frac{1}{2}}\left(\frac{k_{j-1}}{2}D^r_{i-\frac{1}{2},j-\frac{1}{2}} + \frac{k_j}{2}D^r_{i-\frac{1}{2},j+\frac{1}{2}}\right)\left(\frac{u_{i,j} - u_{i-1,j}}{h_{i-1}}\right)$$

$$+ \left(\frac{h_{i-1}}{2}D^z_{i-\frac{1}{2},j+\frac{1}{2}} + \frac{h_i}{2}D^z_{i+\frac{1}{2},j+\frac{1}{2}}\right) r_i \left(\frac{u_{i,j} - u_{i,j+1}}{k_{j+1}}\right)$$

$$+ \left(\frac{h_{i-1}}{2}D^z_{i-\frac{1}{2},j-\frac{1}{2}} + \frac{h_i}{2}D^z_{i+\frac{1}{2},j-\frac{1}{2}}\right) r_i \left(\frac{u_{i,j} - u_{i,j-1}}{k_{j-1}}\right) = 0$$

where

$$D^r_{i-\frac{1}{2},j+\frac{1}{2}} = D^r_{i-\frac{1}{2},j-\frac{1}{2}} = D^r_c$$

$$D^r_{i+\frac{1}{2},j+\frac{1}{2}} = D^r_{i+\frac{1}{2},j-\frac{1}{2}} = D^r_s$$

$$D^z_{i+\frac{1}{2},j+\frac{1}{2}} = D^z_{i+\frac{1}{2},j-\frac{1}{2}} = D^z_s$$

$$D^z_{i-\frac{1}{2},j+\frac{1}{2}} = D^z_{i-\frac{1}{2},j-\frac{1}{2}} = D^z_c$$

Now if D^r and D^z are constants, $h_{i-1} = h_i = h$, and $k_{j-1} = k_j = k$, a second-order correct discretization of the continuity equation at $r = r_{sc}$ is

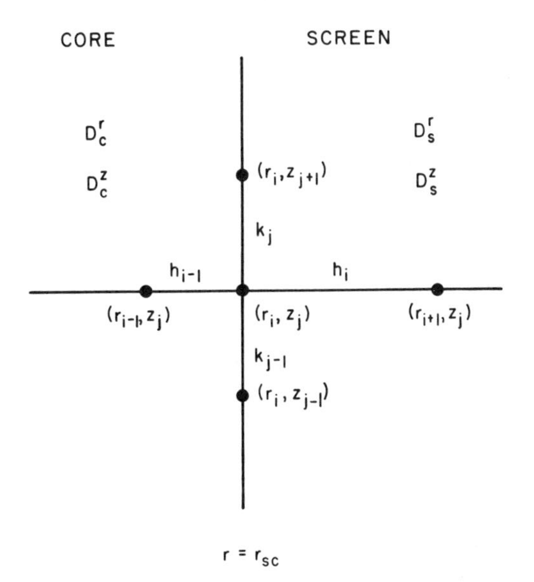

FIGURE 5.3 Grid spacing at core-screen interface of annular bed reactor.

$(r_i = ih,\ z_j = jk)$:

$$\frac{D^r}{h^2}\left[\left(1 + \frac{1}{2i}\right) u_{i+1,j} - 2u_{i,j} + \left(1 - \frac{1}{2i}\right) u_{i-1,j}\right] + \frac{D^z}{k^2}\left[u_{i,j+1} - 2u_{i,j} + u_{i,j-1}\right] = 0$$

Next, we will show that the interface discretization with the conditions stated above simplifies to the previous equation. Since $h_{i-1} = h_i = h$ and $k_{j-1} = k_j = k$, multiply the interface discretization equation by $1/(hkr_i)$ to give

$$\frac{D^r}{h^2 r_i}\left[r_{i+\frac{1}{2}}u_{i+1,j} - (r_{i+\frac{1}{2}} + r_{i-\frac{1}{2}})u_{i,j} + r_{i-\frac{1}{2}}u_{i-1,j}\right] + \frac{D^z}{k^2}\left[u_{i,j+1} - 2u_{i,j} + u_{i,j-1}\right] = 0$$

Notice that

$$\frac{r_{i+\frac{1}{2}}}{r_i} = \frac{(i + \frac{1}{2})h}{ih} = 1 + \frac{1}{2i}$$

$$\frac{r_{i+\frac{1}{2}} + r_{i-\frac{1}{2}}}{r_i} = \frac{(i + \frac{1}{2})h + (i - \frac{1}{2})h}{ih} = 2$$

$$\frac{r_{i-\frac{1}{2}}}{r_i} = \frac{(i - \frac{1}{2})h}{ih} = 1 - \frac{1}{2i}$$

and that with these rearrangements, the previous discretization becomes the second-order correct discretion shown above.

Irregular Boundaries

Dirichlet Condition One method of treating the Dirichlet condition with irregular boundaries is to use unequal mesh spacings. For example, in figure 5.4*a* a vertical mesh spacing from position B of βh and a horizontal mesh spacing of αh would incorporate ∂R into the discretization at the point B.

Another method of treating the boundary condition using a uniform mesh involves selecting a new boundary. Referring to Figure 5.4*a*, given the curve ∂R, one might select the new boundary to pass through position B, that is, (x_B, y_B). Then, a zeroth-degree interpolation would be to take u_B to be $f(x_B, y_B + \beta h)$ or $f(x_B + \alpha h, y_B)$ where $w = f(x, y)$ on ∂R. The replacement of u_B by $f(x_B, y_B + \beta h)$ can be considered as interpolation at B by a polynomial of degree zero with value $f(x_B, y_B + \beta h)$ at $(x_B, y_B + \beta h)$. Hence the term interpolation of degree zero. A more precise approximation is obtained by an interpolation of degree one. A first-degree interpolation using positions u_B and u_C is:

$$\frac{u_B - f(x_B, y_B + \beta h)}{\beta h} = \frac{u_C - u_b}{h}$$

or

$$u_B = \left(\frac{\beta}{\beta + 1}\right)u_C + \left(\frac{1}{\beta + 1}\right)f(x_B, y_B + \beta h) \tag{5.25}$$

Alternatively, we could have interpolated in the x-direction to give

$$u_B = \left(\frac{\alpha}{\alpha + 1}\right)u_A + \left(\frac{1}{\alpha + 1}\right)f(x_B + \alpha h, y_B) \tag{5.26}$$

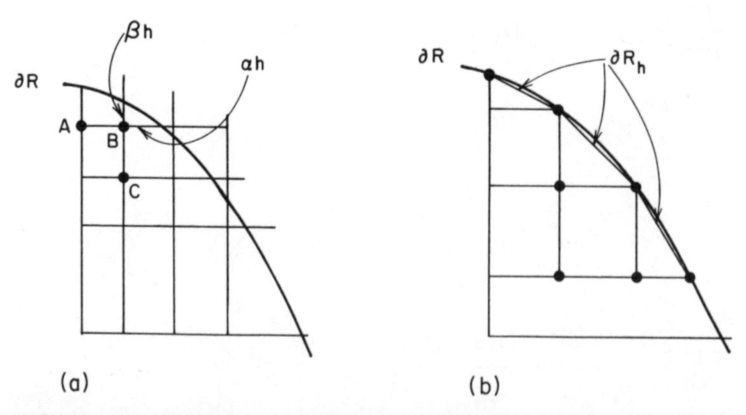

FIGURE 5.4 Irregular boundaries. (*a*) Uniform mesh with interpolation. (*b*) Nonuniform mesh with approximate boundary ∂R_h.

Normal Derivative Conditions. Fortunately, in many practical applications normal derivative conditions occur only along straight lines, e.g., lines of symmetry, and often these lines are parallel to a coordinate axis. However, in the case where the normal derivative condition exists on an irregular boundary, it is suggested that the boundary ∂R be approximated by straight-line segments denoted ∂R_h in Figure 5.4(b). In this situation the use of nonuniform grids is required. To implement the integration method at the boundary ∂R_h, refer to Figure 5.5 during the following analysis. If $b(x_i, y_j) \neq 0$ in (5.14), then $u_{i,j}$ is unknown. The approximation (5.22) can be used for vertical and horizontal portions of the line integral in Figure 5.5, but not on the portion denoted ∂R_h. On ∂R_h the normal to ∂R_h makes an angle θ with the positive x-axis. Thus, ∂R_h must be parameterized by

$$
\begin{aligned}
x &= x_{i+1/2} - \lambda \sin \theta \\
y &= y_{j-1/2} + \lambda \cos \theta
\end{aligned}
\tag{5.27}
$$

and on ∂R_h

$$
\frac{\partial w}{\partial n} = w_x \cos \theta + w_y \sin \theta \tag{5.28}
$$

The portion of the line integral $(x_{i+1/2}, y_{j-1/2})$ to (x_i, y_j) in (5.20) can be written as

$$
-\int_0^{\ell} \{Pw_x \, dy - Pw_y \, dx\}
$$

$$
= -\int_0^{\ell} (Pw_x \cos \theta + Pw_y \sin \theta) \, d\lambda = -\int_0^{\ell} P \frac{\partial w}{\partial n} \, d\lambda
$$

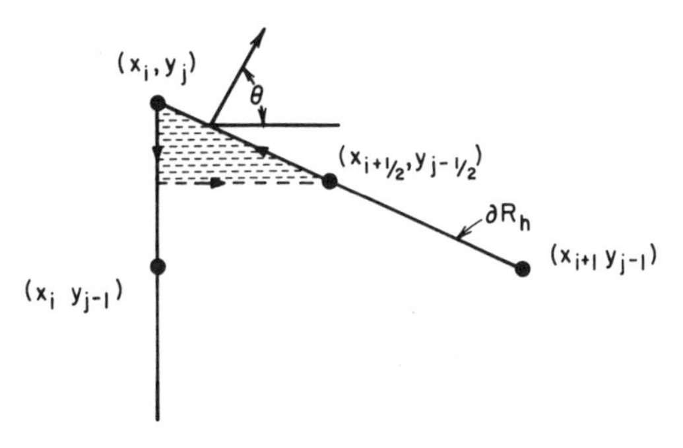

FIGURE 5.5 Boundary point on ∂R_h. Adapted from Richard S. Varga, *Matrix Iterative Analysis*, © 1962, p. 184. Reprinted by permission of Prentice-Hall, Inc., Englewood Cliffs, N. J.

or by using (5.14):

$$-\int_0^{\ell} \{Pw_x\,dy - Pw_y\,dx\} = -\int_0^{\ell} P\left[\frac{c(\lambda) - a(\lambda)w(\lambda)}{b(\lambda)}\right] d\lambda$$

$$\simeq -P_{i,j}\left[\frac{c_{i,j} - a_{i,j}u_{i,j}}{b_{i,j}}\right]\ell \qquad \textbf{(5.29)}$$

where

$$\ell = \frac{1}{2}\sqrt{h_i^2 + k_{j-1}^2} \qquad \text{(path length of integration).}$$

Notice that we have used the boundary condition together with the differential equation to obtain a difference equation for the point (x_i, y_j).

ELLIPTIC PDES—FINITE ELEMENTS

Background

Let us begin by illustrating finite element methods with the following elliptic PDE:

$$\frac{\partial^2 w}{\partial x^2} + \frac{\partial^2 w}{\partial y^2} = -f(x, y), \qquad \text{for } (x, y) \text{ in } R \qquad \textbf{(5.30)}$$

and

$$w(x, y) = 0, \qquad \text{for } (x, y) \text{ on } \partial R \qquad \textbf{(5.31)}$$

Let the bounded domain R with boundary ∂R be the unit square, that is, $0 \leqslant x \leqslant 1$, $0 < y \leqslant 1$. Finite element methods find a piecewise polynomial (pp) approximation, $u(x, y)$, to the solution of (5.30). The pp-approximation can be written as

$$u(x, y) = \sum_{j=1}^{m} \alpha_j \phi_j(x, y) \qquad \textbf{(5.32)}$$

where $\{\phi_j(x, y) | j = 1, \ldots, m\}$ are specified functions that satisfy the boundary conditions and $\{\alpha_j | j = 1, \ldots, m\}$ are as yet unknown constants.

In the collocation method the set $\{\alpha_j | j = 1, \ldots, m\}$ is determined by satisfying the PDE exactly at m points, $\{(x_i, y_i) | i = 1, \ldots, m\}$, the collocation points in the region. The collocation problem for (5.30) using (5.32) as the pp-approximation is given by:

$$A^C \boldsymbol{\alpha} = -\mathbf{f} \qquad \textbf{(5.33)}$$

where

$$A_{ij}^C = \frac{\partial^2 \phi_j}{\partial x^2}(x_i, y_i) + \frac{\partial^2 \phi_j}{\partial y^2}(x_i, y_i)$$

$$\boldsymbol{\alpha} = [\alpha_1, \alpha_2, \ldots, \alpha_m]^T$$

$$\mathbf{f} = [f(x_1, y_1), \ldots, f(x_m, y_m)]^T$$

The solution of (5.33) then yields the vector $\boldsymbol{\alpha}$, which determines the collocation approximation.

To formulate the Galerkin method, first multiply (5.30) by ϕ_i and integrate over the unit square:

$$\iint_R \left(\frac{\partial^2 w}{\partial x^2} + \frac{\partial^2 w}{\partial y^2}\right) \phi_i \, dx \, dy = -\iint_R f(x, y)\phi_i \, dx \, dy$$

$$i = i, \ldots, m \tag{5.34}$$

Green's first identity for a function t is

$$\iint_R \left(\frac{\partial t}{\partial x}\frac{\partial \phi_i}{\partial x} + \frac{\partial t}{\partial y}\frac{\partial \phi_i}{\partial y}\right) dx \, dy$$

$$= -\iint_R \left(\frac{\partial^2 t}{\partial x^2} + \frac{\partial^2 t}{\partial y^2}\right) \phi_i \, dx \, dy + \int_{\partial R} \frac{\partial t}{\partial n} \phi_i \, d\ell \tag{5.35}$$

where

$\frac{\partial}{\partial n}$ = denotes differentiation in the direction of outward normal

ℓ = path of integration for the line integral

Since the functions ϕ_i satisfy the boundary condition, each ϕ_i is zero on ∂R. Therefore, applying Green's first identity to (5.34) gives

$$\iint_R \left(\frac{\partial w}{\partial x}\frac{\partial \phi_i}{\partial x} + \frac{\partial w}{\partial y}\frac{\partial \phi_i}{\partial y}\right) dx \, dy = \iint_R f(x, y)\, \phi_i \, dx \, dy$$

$$i = 1, \ldots, m \tag{5.36}$$

For any two piecewise continuous functions η and ψ denote

$$(\eta, \psi) = \iint_R \eta\psi \, dx \, dy \tag{5.37}$$

Equation (5.36) can then be written as

$$(\nabla w, \nabla \phi_i) = (f, \phi_i), \qquad i = 1, \ldots, m \tag{5.38}$$

where

∇ = gradient operator.

This formulation of (5.30) is called the weak form. The Galerkin method consists in finding $u(x)$ such that

$$(\nabla u, \nabla \phi_i) = (f, \phi_i), \qquad i = 1, \ldots, m \tag{5.39}$$

or in matrix notation,

$$A^G \boldsymbol{\alpha} = \mathbf{g} \tag{5.40}$$

where

$$\begin{aligned} A_{ij}^G &= (\nabla \phi_i, \nabla \phi_j) \\ \boldsymbol{\alpha} &= [\alpha_1, \ldots, \alpha_m]^T \\ \mathbf{g} &= [g_1, \ldots, g_m]^T \\ g_i &= (f, \phi_i) \end{aligned}$$

Next, we discuss each of these methods in further detail.

Collocation

In Chapter 3 we outlined the collocation procedure for BVPs and found that one of the major considerations in implementing the method was the choice of the approximating space. This consideration is equally important when solving PDEs (with the added complication of another spatial direction). The most straightforward generalization of the basis functions from one to two spatial dimensions is obtained by considering tensor products of the basis functions for the one-dimensional space $\mathscr{P}_k^{\nu}(\pi)$ (see Chapter 3). To describe these piecewise polynomial functions let the region R be a rectangle with $a_1 \leqslant x \leqslant b_1$, $a_2 \leqslant y \leqslant b_2$, where $-\infty < a_i \leqslant b_i < \infty$ for $i = 1, 2$. Using this region Birkhoff et al. [5] and later Bramble and Hilbert [6,7] established and generalized interpolation results for tensor products of piecewise Hermite polynomials in two space variables. To describe their results, let

$$\begin{aligned} \pi_1&: \quad a_1 = x_1 < x_2 < \ldots < x_{N_x+1} = b_1 \\ \pi_2&: \quad a_2 = y_1 < y_2 < \ldots < y_{N_y+1} = b_2, \\ h &= \max_{1 \leqslant i \leqslant N_x} h_i = \max_{1 \leqslant i \leqslant N_x} (x_{i+1} - x_i) \\ k &= \max_{1 \leqslant j \leqslant N_y} k_j = \max_{1 \leqslant j \leqslant N_y} (y_{j+1} - y_j) \\ \rho &= \max \{h, k\} \end{aligned} \tag{5.41}$$

be the partitions in the x- and y-directions, and set $\pi = \pi_1 \times \pi_2$. Denote by $\mathscr{P}^2(\pi)$ the set of all real valued piecewise polynomial functions ϕ_i defined on π such that on each subrectangle $[x_i, x_{i+1}] \times [y_j, y_{j+1}]$ of R defined by π, ϕ_i is a polynomial of degree at most 3 in each variable (x or y). Also, each ϕ_i,

$(\partial\phi_i)/(\partial x)$, and $(\partial\phi_i)/(\partial y)$ must be piecewise continuous. A basis for $\mathscr{P}^2$ is the tensor products of the Hermite cubic basis given in Chapter 3 and is

$$\left\{ v_i(x)v_j(y),\ s_i(x)v_j(y),\ v_i(x)s_j(y),\ s_i(x)s_j(y) \right\} \Bigg|_{i=1}^{N_x+1} \Bigg|_{j=1}^{N_y+1} \tag{5.42}$$

where the v's and s's are listed in Table 3.2. If the basis is to satisfy the homogeneous Dirichlet conditions, then it can be written as:

$$\begin{cases} s_i(x)v_j(y),\ \ s_i(x)s_j(y) & i = 1, N_x + 1,\ \ j = 1, \ldots, N_y + 1 \\ v_i(x)s_j(y),\ \ s_i(x)s_j(y) & i = 1, \ldots, N_x + 1,\ \ j = 1, N_y + 1 \\ v_i(x)v_j(y),\ \ s_i(x)v_j(y),\ \ v_i(x)s_j(y), s_i(x)s_j(y), & i = 2, \ldots, N_x,\ \ j = 2, \ldots, N_y \end{cases} \tag{5.43}$$

Using this basis, Prenter and Russell [8] write the pp-approximation as:

$$\begin{aligned} u(x, y) = \sum_{i=1}^{N_x+1} \sum_{j=1}^{N_y+1} \Bigg[& u(x_i, y_j)v_iv_j + \frac{\partial u}{\partial x}(x_i, y_j)s_iv_j \\ & + \frac{\partial u}{\partial y}(x_i, y_j)v_is_j + \frac{\partial^2 u}{\partial x\, \partial y}(x_i, y_j)s_is_j \Bigg] \end{aligned} \tag{5.44}$$

which involves $4(N_x + 1)(N_y + 1)$ unknown coefficients. On each subrectangle $[x_i, x_{i+1}] \times [y_j, y_{j+1}]$ there are four collocation points that are the combination of the two Gaussian points in the x direction, and the two Gaussian points in the y direction, and are:

$$\begin{aligned} \tau_{i,j}^1 &= \left(x_i + \frac{h_i}{2}\left[1 - \frac{1}{\sqrt{3}}\right], y_j + \frac{k_j}{2}\left[1 - \frac{1}{\sqrt{3}}\right] \right) \\ \tau_{i,j}^2 &= \left(x_i + \frac{h_i}{2}\left[1 + \frac{1}{\sqrt{3}}\right], y_j + \frac{k_j}{2}\left[1 - \frac{1}{\sqrt{3}}\right] \right) \\ \tau_{i,j}^3 &= \left(x_i + \frac{h_i}{2}\left[1 - \frac{1}{\sqrt{3}}\right], y_j + \frac{k_j}{2}\left[1 + \frac{1}{\sqrt{3}}\right] \right) \\ \tau_{i,j}^4 &= \left(x_i + \frac{h_i}{2}\left[1 + \frac{1}{\sqrt{3}}\right], y_j + \frac{k_j}{2}\left[1 + \frac{1}{\sqrt{3}}\right] \right) \end{aligned} \tag{5.45}$$

Collocating at these points gives $4N_xN_y$ equations. The remaining $4N_x + 4N_y + 4$ equations required to determine the unknown coefficients are supplied by the boundary conditions [37]. To obtain the boundary equations on the sides $x = a_1$ and $x = b_1$ differentiate the boundary conditions with respect to y. For example, if

$$\frac{\partial u}{\partial x} = y^2 \quad \text{at} \quad x = a_1 \text{ and } x = b_1$$

then (5.46)

$$\frac{\partial^2 u}{\partial x\, \partial y} = 2y \quad \text{at} \quad x = a_1 \text{ and } x = b_1$$

Equation (5.46) applied at $N_y - 1$ boundary nodes $(y_j | j = 2, \ldots, N_y)$ gives:

$$\begin{aligned} \frac{\partial u}{\partial x}(a_1, y_j) &= y_j^2 \\ \frac{\partial^2 u}{\partial x\, \partial y}(a_1, y_j) &= 2y_j \\ \frac{\partial u}{\partial x}(b_1, y_j) &= y_j^2 \\ \frac{\partial^2 u}{\partial x\, \partial y}(b_1, y_j) &= 2y_j \end{aligned} \tag{5.47}$$

or $4N_y - 4$ equations. A similar procedure at $y = a_2$ and $y = b_2$ is followed to give $4N_x - 4$ equations. At each corner both of the above procedures are applied. For example, if

$$u(a_1, a_2) = g(a_1, a_2) \tag{5.48}$$

then

$$\frac{\partial u}{\partial x}(a_1, a_2) = \frac{\partial g}{\partial x}(a_1, a_2)$$

$$\frac{\partial u}{\partial y}(a_1, a_2) = \frac{\partial g}{\partial y}(a_1, a_2)$$

Thus, the four corners supply the final 12 equations necessary to completely specify the unknown coefficients of (5.44).

EXAMPLE 3

Set up the colocation matrix problem for the PDE:

$$\frac{\partial^2 w}{\partial x^2} + \frac{\partial^2 w}{\partial y^2} = \Phi, \qquad 0 \leqslant x \leqslant 1, \quad 0 \leqslant y \leqslant 1$$

with

$$\begin{aligned} w &= 0, &\quad &\text{for } x = 1 \\ w &= 0, &&\text{for } y = 1 \\ \frac{\partial w}{\partial x} &= 0, &&\text{for } x = 0 \\ \frac{\partial w}{\partial y} &= 0, &&\text{for } y = 0 \end{aligned}$$

where Φ is a constant. This PDE could represent the material balance of an isothermal square catalyst pellet with a zero-order reaction or fluid flow in a rectangular duct under the influence of a pressure gradient. Let $N_x = N_y = 2$.

SOLUTION

Using (5.44) as the pp-approximation requires the formulation of 36 equations. Let us begin by constructing the internal boundary node equations (refer to Figure 5.6*a* for node numberings):

$$\frac{\partial w}{\partial x}(1, 2) = 0, \qquad \frac{\partial^2 w}{\partial x\, \partial y}(1, 2) = 0$$

$$w(3, 2) = 0, \qquad \frac{\partial w}{\partial y}(3, 2) = 0$$

$$\frac{\partial w}{\partial y}(2, 1) = 0, \qquad \frac{\partial^2 w}{\partial x\, \partial y}(2, 1) = 0$$

$$w(2, 3) = 0, \qquad \frac{\partial w}{\partial x}(2, 3) = 0$$

where $w(i, j) = w(x_i, y_j)$. At the corners

$$\frac{\partial w}{\partial x}(1, 1) = \frac{\partial w}{\partial y}(1, 1) = \frac{\partial^2 w}{\partial x\, \partial y}(1, 1) = 0$$

$$w(1, 3) = \frac{\partial w}{\partial x}(1, 3) = \frac{\partial^2 w}{\partial x\, \partial y}(1, 3) = 0$$

$$w(3, 1) = \frac{\partial w}{\partial y}(3, 1) = \frac{\partial^2 w}{\partial x\, \partial y}(3, 1) = 0$$

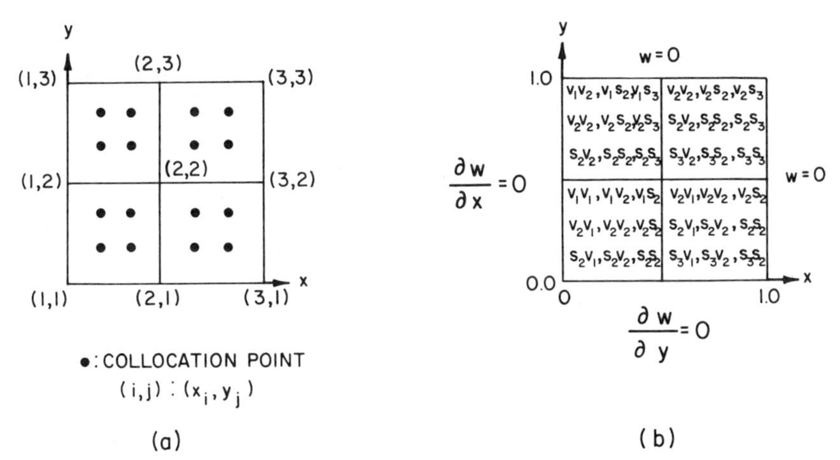

FIGURE 5.6 Grid for Example 3. (*a*) Collocation points. (*b*) Nonvanishing basis functions.

and

$$w(3, 3) = \frac{\partial w}{\partial x}(3, 3) = \frac{\partial w}{\partial y}(3, 3) = 0$$

This leaves 16 equations to be specified. The remaining 16 equations are the collocation equations, four per subrectangle (see Figure 5.6*a*). If the above equations involving the boundary are incorporated in the pp-approximation, then the result is

$$\begin{aligned} u(x, y) = {} & u_{1,1}v_1v_1 + u_{1,2}v_1v_2 + \frac{\partial u_{1,2}}{\partial y} v_1s_2 + \frac{\partial u_{1,3}}{\partial y} v_1s_3 \\ & + u_{2,1}v_2v_1 + \frac{\partial u_{2,1}}{\partial x} s_2v_1 + u_{2,2}v_2v_2 + \frac{\partial u_{2,2}}{\partial x} s_2v_2 \\ & + \frac{\partial u_{2,2}}{\partial y} v_2s_2 + \frac{\partial^2 u_{2,2}}{\partial x\, \partial y} s_2s_2 + \frac{\partial u_{2,3}}{\partial y} v_2s_3 + \frac{\partial^2 u_{2,3}}{\partial x\, \partial y} s_2s_3 \\ & + \frac{\partial u_{3,1}}{\partial x} s_3v_1 + \frac{\partial u_{3,2}}{\partial x} s_3v_2 + \frac{\partial^2 u_{3,2}}{\partial x\, \partial y} s_3s_2 + \frac{\partial^2 u_{3,3}}{\partial x\, \partial y} s_3s_3 \end{aligned}$$

where

$$u_{i,j} = u(x_i, y_j)$$

The pp-approximation is then used to collocate at the 16 collocation points. Since the basis is local, various terms of the above pp-approximation can be zero at a given collocation point. The nonvanishing terms of the pp-approximation are given in the appropriate subrectangle in Figure 5.6*b*. Collocating at the 16 collocation points using the pp-approximation listed above gives the following matrix problem:

$$\mathbf{A}^C\mathbf{a} = \mathbf{\Phi 1}$$

where

$$\mathbf{1} = [1, \ldots, 1]^T$$

$$\begin{aligned} \mathbf{a} = \Bigg[& u_{1,1}, u_{1,2}, \frac{\partial u_{1,2}}{\partial y}, \frac{\partial u_{1,3}}{\partial y}, u_{2,1}, \frac{\partial u_{2,1}}{\partial x}, u_{2,2} \\ & \frac{\partial u_{2,2}}{\partial x}, \frac{\partial u_{2,2}}{\partial y}, \frac{\partial^2 u_{2,2}}{\partial x\, \partial y}, \frac{\partial u_{2,3}}{\partial y}, \frac{\partial^2 u_{2,3}}{\partial x\, \partial y} \\ & \frac{\partial u_{3,1}}{\partial x}, \frac{\partial u_{3,2}}{\partial x}, \frac{\partial^2 u_{3,2}}{\partial x\, \partial y}, \frac{\partial^2 u_{3,3}}{\partial x\, \partial y} \Bigg]^T \end{aligned}$$

$$\nabla^2_{ijs}\psi = \frac{\partial^2}{\partial x^2}\psi(\tau^s_{i,j}) + \frac{\partial^2}{\partial y^2}\psi(\tau^s_{i,j}) \qquad \text{(for any function } \psi\text{)}$$

and for the matrix A^C,

$$
A^C = \begin{bmatrix}
\nabla^2_{111}v_1v_1 & \nabla^2_{111}v_1v_2 & \nabla^2_{111}v_1s_2 & \nabla^2_{111}v_2v_1 & \nabla^2_{111}s_2v_1 & \nabla^2_{111}v_2v_2 & \nabla^2_{111}s_2v_2 & \nabla^2_{111}v_2s_2 & \nabla^2_{111}s_2s_2 & & & \nabla^2_{111}s_3v_1 & & & \\
 & \vdots & & & & \vdots & & & & & & \vdots & & & \\
\nabla^2_{114}v_1v_1 & \nabla^2_{114}v_1v_2 & \nabla^2_{114}v_1s_2 & \nabla^2_{114}v_2v_1 & \nabla^2_{114}s_2v_1 & \nabla^2_{114}v_2v_2 & \nabla^2_{114}s_2v_2 & \nabla^2_{114}v_2s_2 & \nabla^2_{114}s_2s_2 & & & \nabla^2_{114}s_3v_1 & & & \\
 & & & & & \vdots & & & & & & & & & \\
 & & & & & \nabla^2_{221}v_2v_2 & \nabla^2_{221}s_2v_2 & \nabla^2_{221}v_2s_2 & \nabla^2_{221}s_2s_2 & \nabla^2_{221}v_2s_3 & \nabla^2_{221}s_2s_3 & & \nabla^2_{221}s_3v_2 & \nabla^2_{221}s_3s_2 & \nabla^2_{221}s_3s_3 \\
 & & & & & & & & \vdots & & & & & \vdots & \\
 & & & & & \nabla^2_{224}v_2v_2 & \nabla^2_{224}s_2v_2 & \nabla^2_{224}v_2s_2 & \nabla^2_{224}s_2s_2 & \nabla^2_{224}v_2s_3 & \nabla^2_{224}s_2s_3 & & \nabla^2_{224}s_3v_2 & \nabla^2_{224}s_3s_2 & \nabla^2_{224}s_3s_3
\end{bmatrix}
$$

The solution of this matrix problem yields the vector **a**, which specifies the values of the function and its derivatives at the grid points.

Thus far we have discussed the construction of the collocation matrix problem using the tensor products of the Hermite cubic basis for a linear PDE. If one were to solve a nonlinear PDE using this basis, the procedure would be the same as outlined above, but the ensuing matrix problem would be nonlinear.

In Chapter 3 we saw that the expected error in the pp-approximation when solving BVPs for ODEs was dependent upon the choice of the approximating space, and for the Hermite cubic space, was $0(h^4)$. This analysis can be extended to PDEs in two spatial dimensions with the result that [8]:

$$|u(x, y) - w(x, y)| = 0(\rho^4)$$

Next, consider the tensor product basis for $\mathscr{P}^{\nu}_{k_x}(\pi_1) \times \mathscr{P}^{\nu}_{k_y}(\pi_2)$ where π_1 and π_2 are given in (5.41), k_x is the order of the one-dimensional approximating space in the x-direction, and k_y is the order of the one-dimensional approximating space in the y-direction. A basis for this space is given by the tensor products of the B-splines as:

$$B_i^x(x) B_j^y(y) \Big|_{i=1}^{\text{DIMX}} \Big|_{j=1}^{\text{DIMY}} \tag{5.49}$$

where

$$\begin{aligned} B_i^x(x) &= \text{B-spline in the } x\text{-direction of order } k_x \\ B_j^y(y) &= \text{B-spline in the } y\text{-direction or order } k_y \\ \text{DIMX} &= \text{dimension of } \mathscr{P}^{\nu}_{k_x} \\ \text{DIMY} &= \text{dimension of } \mathscr{P}^{\nu}_{k_y} \end{aligned}$$

The pp-approximation for this space is given by

$$u(x, y) = \sum_{i=1}^{\text{DIMX}} \sum_{j=1}^{\text{DIMY}} \alpha_{i,j} B_i^x(x) B_j^y(y) \tag{5.50}$$

where $\alpha_{i,j}$ are constants, with the result that

$$|u(x, y) - w(x, y)| = 0(\rho^{\gamma}) \tag{5.51}$$

where

$$\gamma = \min\{k_x, k_y\}$$

Galerkin

The literature on the use of Galerkin-type methods for the solution of elliptic PDEs is rather extensive and is continually expanding. The reason for this growth in use is related to the ease with which the method accommodates complicated geometries. First, we will discuss the method for rectangles, and then treat the analysis for irregular geometries.

Consider a region R that is a rectangle with $a_1 \leqslant x \leqslant b_1$, $a_2 \leqslant y \leqslant b_2$, with $-\infty < a_i \leqslant b_i < \infty$ for $i = 1, 2$. A basis for the simplest approximating space is obtained from the tensor products of the one-dimensional basis of the space $\mathscr{P}_2^1(\pi)$, i.e., the piecewise linears. If the mesh spacings in x and y are given by π_1 and π_2 of (5.41), then the tensor product basis functions $\omega_{i,j}(x, y)$ are given by

$$\omega_{i,j} = \begin{cases} \left[\dfrac{x - x_{i-1}}{h_{i-1}}\right]\left[\dfrac{y - y_{j-1}}{k_{j-1}}\right], & x_{i-1} \leqslant x \leqslant x_i, \quad y_{j-1} \leqslant y \leqslant y_j \\ \left[\dfrac{x - x_{i-1}}{h_{i-1}}\right]\left[\dfrac{y_{j+1} - y}{k_j}\right], & x_{i-1} \leqslant x \leqslant x_i, \quad y_j \leqslant y \leqslant y_{j+1} \\ \left[\dfrac{x_{i+1} - x}{h_i}\right]\left[\dfrac{y - y_{j-1}}{k_{j-1}}\right], & x_i \leqslant x \leqslant x_{i+1}, \quad y_{j-1} \leqslant y \leqslant y_j \\ \left[\dfrac{x_{i+1} - x}{h_i}\right]\left[\dfrac{y_{j+1} - y}{k_j}\right], & x_i \leqslant x \leqslant x_{i+1}, \quad y_j \leqslant y \leqslant y_{j+1} \end{cases} \tag{5.52}$$

with a pp-approximation of

$$u(x, y) = \sum_{i=1}^{Nx+1} \sum_{j=1}^{Ny+1} u(x_i, y_j)\omega_{i,j} \tag{5.53}$$

Therefore, there are $(N_x + 1)(N_y + 1)$ unknown constants $u(x_i, y_j)$, each associated with a given basis function $\omega_{i,j}$. Figure 5.7 illustrates the basis function $\omega_{i,j}$, from now on called a bilinear basis function.

EXAMPLE 4

Solve (5.30) with $f(x, y) = 1$ using the bilinear basis with $N_x = N_y = 2$.

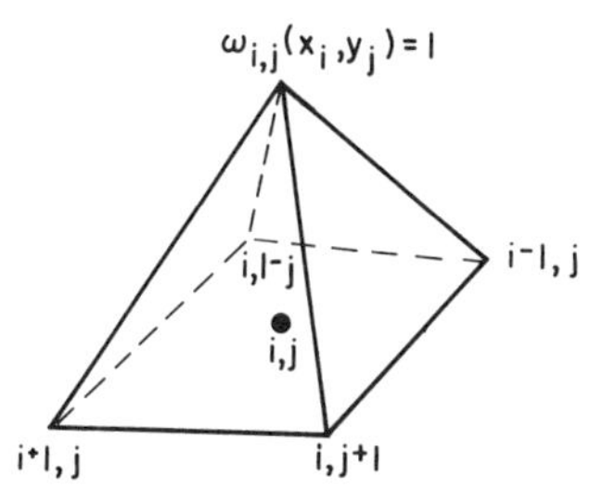

FIGURE 5.7 Bilinear basis function.

SOLUTION

The PDE is

$$\frac{\partial^2 w}{\partial x^2} + \frac{\partial^2 w}{\partial y^2} = -1, \qquad 0 \leq x \leq 1, \quad 0 \leq y \leq 1$$

with

$$w(x, y) = 0 \qquad \text{on the boundary}$$

The weak form of the PDE is

$$\iint_R \left(\frac{\partial w}{\partial x} \frac{\partial \phi_i}{\partial x} + \frac{\partial w}{\partial y} \frac{\partial \phi_i}{\partial y} \right) dx\, dy = \iint_R \phi_i \, dx\, dy$$

where each ϕ_i satisfies the boundary conditions. Using (5.53) as the pp-approximation gives

$$u(x, y) = \sum_{i=1}^{3} \sum_{j=1}^{3} u(x_i, y_j)\omega_{i,j}$$

Let $h_i = k_j = h = 0.5$ as shown in Figure 5.8, and number each of the subrectangles, which from now on will be called elements. Since each $\omega_{i,j}$ must satisfy the boundary conditions,

$$\omega_{1,1} = \omega_{1,2} = \omega_{1,3} = \omega_{2,1} = \omega_{3,1} = \omega_{2,3} = \omega_{3,2} = \omega_{3,3} = 0$$

leaving the pp-approximation to be

$$u(x, y) = u(x_2, y_2)\omega_{2,2} = u_2\omega_2$$

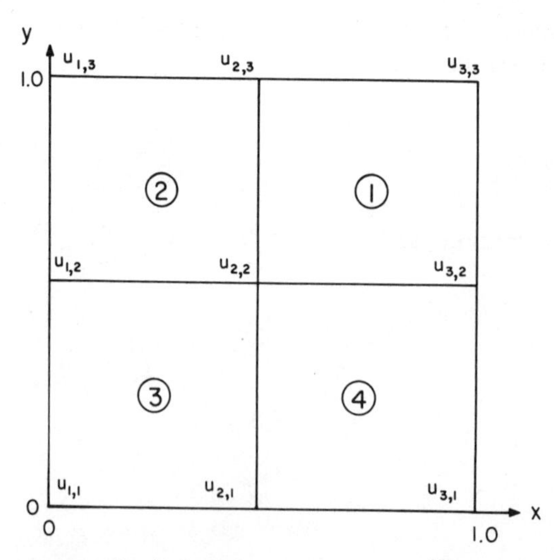

FIGURE 5.8 Grid for Example 4. ⓘ **= element *i*.**

Therefore, upon substituting $u(x, y)$ for $w(x, y)$, the weak form of the PDE becomes

$$\iint_R \left(u_2 \frac{\partial \omega_2}{\partial x} \frac{\partial \omega_2}{\partial x} + u_2 \frac{\partial \omega_2}{\partial y} \frac{\partial \omega_2}{\partial y} \right) dx\, dy = \iint_R \omega_2\, dx\, dy$$

or

$$A_{22} u_2 = g_2$$

where

$$A_{22} = \iint_R \left(\frac{\partial \omega_2}{\partial x} \frac{\partial \omega_2}{\partial x} + \frac{\partial \omega_2}{\partial y} \frac{\partial \omega_2}{\partial y} \right) dx\, dy$$

$$g_2 = \iint_R \omega_2\, dx\, dy$$

This equation can be solved on a single element e_i as

$$A_{22}^{e_i} u_2 = g_2^{e_i}, \qquad e_i = 1, \ldots, 4$$

and then summed over all the elements to give

$$A_{22} u_2 = \sum_{e_i=1}^{4} A_{22}^{e_i} u_2 = \sum_{e_i=1}^{4} g_2^{e_i} = g_2$$

In element 1:

$$u(x, y) = \frac{u_2}{h^2} (1 - x)(1 - y), \qquad 0.5 \leq x \leq 1, \quad 0.5 \leq y \leq 1$$

and

$$\omega_2 = \frac{1}{h^2} (1 - x)(1 - y)$$

Thus

$$A_{22}^1 = \frac{1}{h^4} \int_{0.5}^1 \int_{0.5}^1 [(1 - y)^2 + (1 - x)^2]\, dx\, dy = \frac{2}{3} \qquad (h = 0.5)$$

and

$$g_2^1 = \frac{1}{h^2} \int_{0.5}^1 \int_{0.5}^1 (1 - x)(1 - y)\, dx\, dy = \frac{h^2}{4}$$

For element 2:

$$u(x, y) = \frac{u_2}{h^2} (1 - y)x, \qquad 0 \leq x \leq 0.5, \quad 0.5 \leq y \leq 1.0$$

and

$$\omega_2 = \frac{1}{h^2} x(1 - y)$$

giving

$$A_{22}^2 = \tfrac{2}{3}$$

and

$$g_2^2 = \frac{h^2}{4}$$

The results for each element are

Element	$A_{22}^{e_i}$	$g_2^{e_i}$
1	$\frac{2}{3}$	$\frac{h^2}{4}$
2	$\frac{2}{3}$	$\frac{h^2}{4}$
3	$\frac{2}{3}$	$\frac{h^2}{4}$
4	$\frac{2}{3}$	$\frac{h^2}{4}$

Thus, the solution is given by the sum of these results and is

$$u_2 = \tfrac{3}{8} h^2 = 0.09375$$

In the previous example we saw how the weak form of the PDE could be solved element by element. When using the bilinear basis the expected error in the pp-approximation is

$$|u(x, y) - w(x, y)| = 0(\rho^2) \tag{5.54}$$

where ρ is given in (5.41). As with ODEs, to increase the order of accuracy, the order of the tensor product basis functions must be increased, for example, the tensor product basis using Hermite cubics given an error of $0(\rho^4)$. To illustrate the formulation of the Galerkin method using higher-order basis functions, let the pp-approximation be given by (5.50) and reconsider (5.30) as the elliptic PDE. Equation (5.39) becomes

$$\left(\nabla \sum_{i=1}^{\text{DIMX}} \sum_{j=1}^{\text{DIMY}} \alpha_{i,j} B_i^x(x) B_j^y(y),\ \nabla B_m^x(x) B_n^y(y)\right) = \left(f,\ B_m^x(x) B_n^y(y)\right) \tag{5.55}$$

$$m = 1, \ldots, \text{DIMX}, \qquad n = 1, \ldots, \text{DIMY}$$

In matrix notation (5.55) is

$$\mathbf{A}\boldsymbol{\alpha} = \mathbf{g} \tag{5.56}$$

where

$$\boldsymbol{\alpha} = [\tilde{\boldsymbol{\alpha}}_1, \ldots, \tilde{\boldsymbol{\alpha}}_{\mathrm{DIMX}}]^T$$

$$\tilde{\boldsymbol{\alpha}}_i = [\alpha_{i,1}, \ldots, \alpha_{i,\mathrm{DIMY}}]^T$$

$$\mathbf{g} = [\tilde{\mathbf{g}}_1, \ldots, \tilde{\mathbf{g}}_2]^T$$

$$\tilde{\mathbf{g}}_i = [(f, B_i^x(x)B_1^y(y)), \ldots, (f, B_i^x(x)B_{\mathrm{DIMY}}^y(y))]^T$$

$$A_{p,q} = (\nabla B_i^x(x)B_j^y(y), \nabla B_m^x(x)B_n^y(y))$$

$$p = \mathrm{DIMY}\,(m - 1) + n \qquad (1 \leqslant p \leqslant \mathrm{DIMX} \times \mathrm{DIMY})$$

$$q = \mathrm{DIMY}\,(i - 1) + j \qquad (1 \leqslant q \leqslant \mathrm{DIMX} \times \mathrm{DIMY})$$

Equation (5.56) can be solved element by element as

$$\sum_{e_i=1}^{\text{No. of elements}} A_{pq}^{e_i}\alpha_q = \sum_{e_i=1}^{\text{No. of elements}} g_p^{e_i} \tag{5.57}$$

The solution of (5.56) or (5.57) gives the vector $\boldsymbol{\alpha}$, which specifies the pp-approximation $u(x, y)$ with an error given by (5.51).

Another way of formulating the Galerkin solution to elliptic problems is that first proposed by Courant [9]. consider a general plane polygonal region R with boundary ∂R. When the region R is not a rectangular parallelepiped, a rectangular grid does not approximate R and especially ∂R as well as a triangular grid, i.e., covering the region R with a finite number of arbitrary triangles. This point is illustrated in Figure 5.9. Therefore, if the Galerkin method can be formulated with triangular elements, irregular regions can be handled through the use of triangulation. Courant developed the method for Dirichlet-type boundary conditions and used the space of continuous functions that are linear polynomials on each triangle. To illustrate this method consider (5.30) with the pp-approximation (5.32). If there are TN vertices not on ∂R in the triangulation, then (5.32) becomes

$$u(x, y) = \sum_{s=1}^{TN} \alpha_s \phi_s(x, y) \tag{5.58}$$

Given a specific vertex $s = \ell$, $\alpha_\ell = u(x_\ell, y_\ell)$ with an associated basis function $\phi_\ell(x, y)$. Figure 5.10*a* shows the vertex (x_ℓ, y_ℓ) and the triangular elements that contain it, while Figure 5.10*b* illustrates the associated basis function. The weak form of (5.30) is

$$\iint_R \left(\frac{\partial u}{\partial x}\frac{\partial \phi_s}{\partial x} + \frac{\partial u}{\partial y}\frac{\partial \phi_s}{\partial y} \right) dx\,dy = \iint_R f(x, y)\phi_s\,dx\,dy$$

$$s = 1, \ldots, TN \tag{5.59}$$

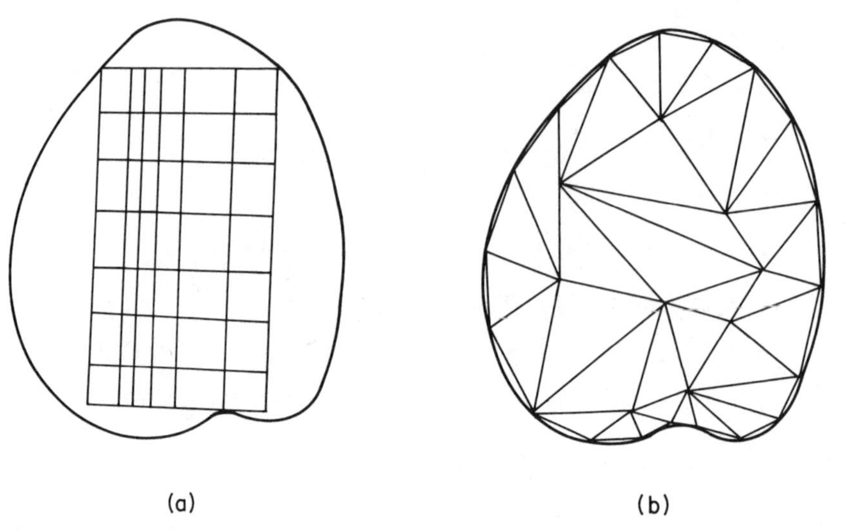

FIGURE 5.9 Grids on a polygonal region. (*a*) Rectangular grid. (*b*) Triangular grid.

or in matrix notation

$$\mathbf{A}\boldsymbol{\alpha} = \mathbf{g} \tag{5.60}$$

where

$$A_{sq} = \iint_R \left[\frac{\partial \phi_s}{\partial x} \frac{\partial \phi_q}{\partial x} + \frac{\partial \phi_s}{\partial y} \frac{\partial \phi_q}{\partial y} \right] dx\, dy$$

$$\boldsymbol{\alpha} = [\alpha_1, \ldots, \alpha_{TN}]^T$$

$$\mathbf{g} = \left[\iint_R f(x, y)\phi_1\, dx\, dy, \ldots, \iint_R f(x, y)\phi_{TN}\, dx\, dy \right]^T$$

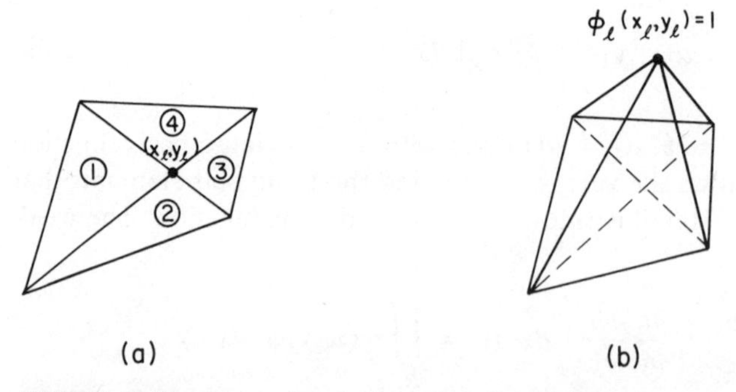

FIGURE 5.10 Linear basis function for triangular elements. (*a*) Vertex (x_ℓ, y_ℓ). (*b*) Basis function ϕ_ℓ.

Equation (5.60) can be solved element by element (triangle by triangle) and summed to give

$$\sum_{e_i} A_{sq}^{e_i} \alpha_q = \sum_{e_i} g_s^{e_i}$$

$$s = 1, \ldots, TN, \qquad q = 1, \ldots, TN \tag{5.61}$$

Since the PDE can be solved element by element, we need only discuss the formulation of the basis functions on a single triangle. To illustrate this formulation, first consider a general triangle with vertices (x_i, y_i), $i = 1, 2, 3$. A linear interpolation $P_1(x, y)$ of a function $C(x, y)$ over the triangle is given by [10]:

$$P_1(x, y) = \sum_{i=1}^{3} a_i(x, y) C(x_i, y_i) \tag{5.62}$$

where

$$a_1(x, y) = \psi(\tau_{23} + \eta_{23}x - \xi_{23}y)$$

$$a_2(x, y) = \psi(\tau_{31} + \eta_{31}x - \xi_{31}y)$$

$$a_3(x, y) = \psi(\tau_{12} + \eta_{12}x - \xi_{12}y)$$

$$\psi = (\text{twice the area of the triangle})^{-1}$$

$$\tau_{ij} = x_i y_j - x_j y_i$$

$$\xi_{ij} = x_i - x_j$$

$$\eta_{ij} = y_i - y_j$$

To construct the basis function ϕ_ℓ associated with the vertex (x_ℓ, y_ℓ) on a single triangle set $(x_\ell, y_\ell) = (x_1, y_1)$ in (5.62). Also, since $\phi_\ell(x_\ell, y_\ell) = 1$ and ϕ_ℓ is zero at all other vertices set $C(x_1, y_1) = 1$, $C(x_2, y_2) = 0$ and $C(x_3, y_3) = 0$ in (5.62). With these substitutions, $\phi_\ell = P_1(x, y) = a_1(x, y)$. We illustrate this procedure in the following example.

EXAMPLE 5

Solve the problem given in Example 3 with the triangulation shown in Figure 5.11.

SOLUTION

From the boundary conditions

$$u_{1,1} = u_{2,1} = u_{3,1} = u_{3,2} = u_{3,3} = u_{2,3} = u_{1,3} = u_{1,2} = 0$$

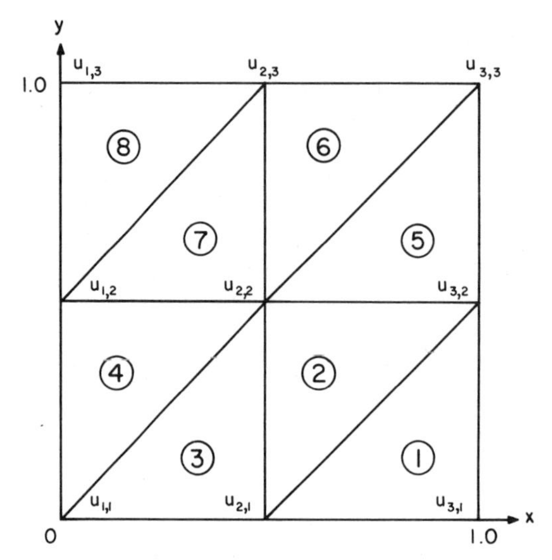

FIGURE 5.11 Triangulation for Example 5. (i) = element *i*

Therefore, the only nonzero vertex is $u_{2,2}$, which is common to elements 2, 3, 4, 5, 6, and 7, and the pp-approximation is given by

$$u(x, y) = u_{2,2}\phi_2(x, y) = u_2\phi_2$$

Equation (5.61) becomes

$$\sum_{e_i=2}^{7} A_{22}^{e_i} u_2 = \sum_{e_i=2}^{7} g_2^{e_i}$$

where

$$A_{22}^{e_i} = \iint\limits_{\substack{\text{Triangle}\\ e_i}} \left(\frac{\partial \phi_2}{\partial x}\frac{\partial \phi_2}{\partial x} + \frac{\partial \phi_2}{\partial y}\frac{\partial \phi_2}{\partial y} \right) dx\, dy$$

$$g_2^{e_i} = \iint\limits_{\substack{\text{Triangle}\\ e_i}} \phi_2\, dx\, dy$$

The basis function $\phi_2^{e_i}$ can be constructed using (5.62) with $(x_1, y_1) = (0.5, 0.5)$ giving

$$\phi_2^{e_i} = \psi(\tau_{23}^{e_i} + \eta_{23}^{e_i} x - \xi_{23}^{e_i} y)$$

Thus,

$$\frac{\partial \phi_2^{e_i}}{\partial x} = \psi \eta_{23}^{e_i}$$

$$\frac{\partial \phi_2^{e_i}}{\partial y} = -\psi \xi_{23}^{e_i}$$

$$A_{22}^{e_i} = \iint\limits_{e_i} \psi^2[(\eta_{23}^{e_i})^2 + (\xi_{23}^{e_i})^2]\, dx\, dy$$

and

$$g_2^{e_i} = \iint_{e_i} \psi(\tau_{23}^{e_i} + \eta_{23}^{e_i}x - \xi_{23}^{e_i}y)\, dx\, dy$$

For element 2 we have the vertices

$$(x_1, y_1) = (0.5, 0.5)$$

$$(x_2, y_2) = (1, 0.5)$$

$$(x_3, y_3) = (0.5, 0)$$

and

$$\psi = \frac{1}{0.25}$$

$$\tau_{23} = (1)(0) - (0.5)(0.5) = -0.25$$

$$\xi_{23} = 1 - 0.5 = 0.5$$

$$\eta_{23} = 0.5$$

$$A_{22}^2 = \iint (0.25)^{-2}[(0.5)^2 + (0.5)^2]\, dx\, dy = 1$$

$$g_2^2 = \iint \frac{1}{(0.25)}[-0.25 + 0.5x - 0.5y]\, dx\, dy = \frac{0.25}{6}$$

Likewise, the results for other elements are

Element	$A_{22}^{e_i}$	$g_2^{e_i}$
2	1.0	$\frac{0.25}{6}$
3	0.5	$\frac{0.25}{6}$
4	0.5	$\frac{0.25}{6}$
5	0.5	$\frac{0.25}{6}$
6	0.5	$\frac{0.25}{6}$
7	1.0	$\frac{0.25}{6}$
Total	4.0	0.25

which gives

$$u_2 = 0.0625$$

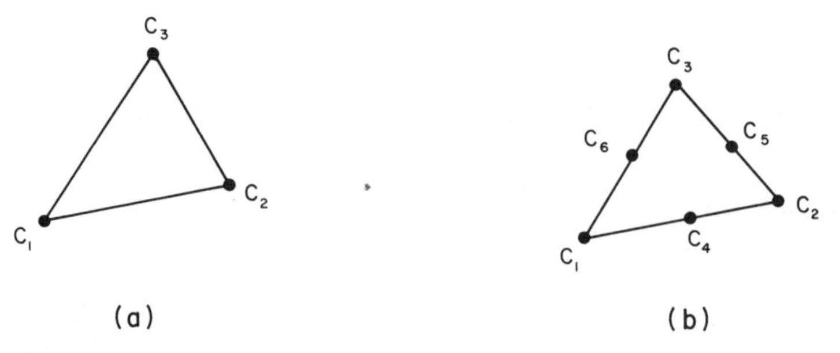

FIGURE 5.12 Node positions for triangular elements. (*a*) Linear basis. (*b*) Quadratic basis: $C_l = C(x_l, y_l)$.

The expected error in the pp-approximation using triangular elements with linear basis functions is $0(h^2)$ [11], where h denotes the length of the largest side of any triangle. As with rectangular elements, to obtain higher-order accuracy, higher-order basis functions must be used. If quadratic functions are used to interpolate a function, $C(x, y)$, over a triangular element, then the interpolation is given by [10]:

$$P_2(x, y) = \sum_{i=1}^{6} b_i(x, y)C(x, y) \tag{5.63}$$

where

$$b_j(x, y) = a_j(x, y)[2a_j(x, y) - 1], \qquad j = 1, 2, 3$$

$$b_4(x, y) = 4a_1(x, y)a_2(x, y)$$

$$b_5(x, y) = 4a_1(x, y)a_3(x, y)$$

$$b_6(x, y) = 4a_2(x, y)a_3(x, y)$$

and the $a_i(x, y)$'s are given in (5.62). Notice that the linear interpolation (5.62) requires three values of $C(x, y)$ while the quadratic interpolation (5.63) requires six. The positions of these values for the appropriate interpolations are shown in Figure 5.12. Interpolations of higher order have also been derived, and good presentations of these bases are given in [10] and [12].

Now, consider the problem of constructing a set of basis functions for an irregular region with a curved boundary. The simplest way to approximate the curved boundary is to construct the triangulation such that the boundary is approximated by the straight-line segements of the triangles adjacent to the boundary. This approximation is illustrated in Figure 5.9*b*. An alternative procedure is to allow the triangles adjacent to the boundary to have a curved side that is part of the boundary. A transformation of the coordinate system can then restore the elements to the standard triangular shape, and the PDE solved as previously outlined. If the same order polynomial is chosen for the coordinate change as for the basis functions, then this method of incorporating the curved boundary is called the isoparametric method [10–12]. To outline the procedure, consider a triangle with one curved edge that arises at a boundary as shown in Figure 5.13*a*. The simplest polynomial able to describe the curved side of the triangular element is a quadratic. Therefore, specify the basis functions for the triangle in the λ_1-λ_2 plane to be quadratics. These basis functions are completely specified by their values at the six nodes shown in Figure 5.13*b*. Thus the isoparametric method maps the six nodes in the *x*-*y* plane onto the λ_1-λ_2 plane. The PDE is solved in this coordinate system, giving $u(\lambda_1, \lambda_2)$, which can be transformed to $u(x, y)$.

PARABOLIC PDES IN TWO SPACE VARIABLES

In Chapter 4 we treated finite difference and finite element methods for solving parabolic PDEs that involved one space variable and time. Next, we extend the discussion to include two spatial dimensions.

Method of Lines

Consider the parabolic PDE

$$\frac{\partial w}{\partial t} = D\left[\frac{\partial^2 w}{\partial x^2} + \frac{\partial^2 w}{\partial y^2}\right]$$

$$0 \leqslant t, \qquad 0 \leqslant x \leqslant 1, \qquad 0 \leqslant y \leqslant 1 \tag{5.64}$$

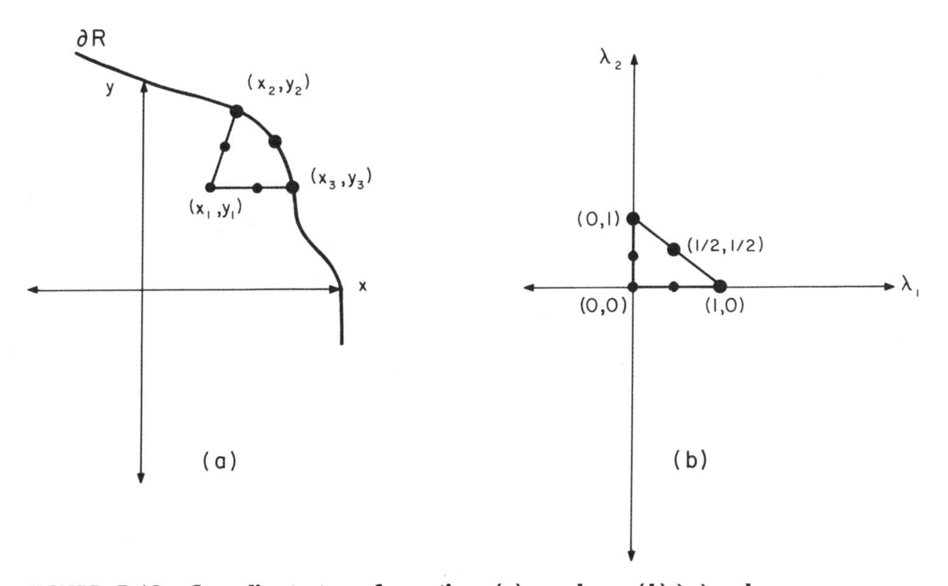

FIGURE 5.13 Coordinate transformation. (*a*) *xy*-plane. (*b*) $\lambda_1\lambda_2$-plane.

with D = constant. Discretize the spatial derivatives in (5.64) using finite differences to obtain the following system of ordinary differential equations:

$$\frac{\partial u_{i,j}}{\partial t} = \frac{D}{(\Delta x)^2}[u_{i+1,j} - 2u_{i,j} + u_{i-1,j}] + \frac{D}{(\Delta y)^2}[u_{i,j+1} - 2u_{i,j} + u_{i,j-1}] \quad (5.65)$$

where

$$u_{i,j} \simeq w(x_i, y_j)$$

$$x_i = i\,\Delta x$$

$$y_j = j\,\Delta y$$

Equation (5.65) is the two-dimensional analog of (4.6) and can be solved in a similar manner. To complete the formulation requires knowledge of the subsidiary conditions. The parabolic PDE (5.64) requires boundary conditions at $x = 0$, $x = 1$, $y = 0$, and $y = 1$, and an initial condition at $t = 0$. As with the MOL in one spatial dimension, the two-dimensional problem incorporates the boundary conditions into the spatial discretizations while the initial condition is used to start the IVP.

Alternatively, (5.64) could be discretized using Galerkin's method or by collocation. For example, if (5.32) is used as the pp-approximation, then the collocation MOL discretization is

$$\sum_{j=1}^{m} \frac{\partial \alpha_j}{\partial t}\,\phi_j(x_i, y_i) = D \sum_{j=1}^{m} \alpha_j \left[\frac{\partial^2}{\partial x^2}\,\phi_j(x_i, y_i) + \frac{\partial^2}{\partial y^2}\,\phi_j(x_i, y_i)\right], \quad (5.66)$$

$$i = 1, \ldots, m$$

where (x_i, y_i) designates the position of the ith collocation point. Since the MOL was discussed in detail in Chapter 4 and since the multidimensional analogs are straightforward extensions of the one-dimensional cases, no rigorous presentation of this technique will be given.

Alternating Direction Implicit Methods

Discretize (5.65) in time using Euler's method to give

$$u_{i,j}^{n} = \left[\frac{D\,\Delta t}{(\Delta x)^2}\right]\left[u_{i+1,j}^{n} + u_{i-1,j}^{n}\right] + \left[\frac{D\,\Delta t}{(\Delta y)^2}\right]\left[u_{i,j+1}^{n} + u_{i,j-1}^{n}\right] + u_{i,j}^{n}\left[1 - \frac{2D\,\Delta t}{(\Delta x)^2} - \frac{2D\,\Delta t}{(\Delta y)^2}\right] \quad (5.67)$$

where

$$u_{i,j}^{n} \simeq w(x_i, y_j, t_n)$$

$$t_n = n\,\Delta t$$

For stability

$$D\,\Delta t\left[\frac{1}{(\Delta x)^2}+\frac{1}{(\Delta y)^2}\right]<\frac{1}{2} \tag{5.68}$$

If $\Delta x = \Delta y$, then (5.68) becomes

$$\frac{D\,\Delta t}{(\Delta x)^2}\leqslant\frac{1}{4} \tag{5.69}$$

which says that the restriction on the time step-size is half as large as the one-dimensional analog. Thus the stable time step-size decreases with increasing dimensionality. Because of the poor stability properties common to explicit difference methods, they are rarely used to solve multidimensional problems. Inplicit methods with their superior stability properties could be used instead of explicit formulas, but the resulting matrix problems are not easily solved. Another approach to the solution of multidimensional problems is to use alternating direction implicit (ADI) methods, which are two-step methods involving the solution of tridiagonal sets of equations (using finite difference discretizations) along lines parallel to the x-y axes at the first-second steps, respectively.

Consider (5.64) with $D = 1$ where the region to be examined in (x, y, t) space is covered by a rectilinear grid with sides parallel to the axes, and $h = \Delta x = \Delta y$. The grid points (x_i, y_j, t_n) given by $x = ih$, $y = jh$, and $t = n\,\Delta t$, and $u^n_{i,j}$ is the function satisfying the finite difference equation at the grid points. Define

$$\begin{aligned}\delta_x u^n_{i,j} &= u^n_{i+1/2,j} - u^n_{i-1/2,j}\\ \delta_y u^n_{i,j} &= u^n_{i,j+1/2} - u^n_{i,j-1/2}\\ \tau &= \frac{\Delta t}{h^2}\end{aligned} \tag{5.70}$$

Essentially, the principle is to employ two difference equations that are used in turn over successive time-steps of $\Delta t/2$. The first equation is implicit in the x-direction, while the second is implicit in the y-direction. Thus, if $\tilde{u}_{i,j}$ is an intermediate value at the end of the first time-step, then

$$\begin{aligned}\tilde{u}_{i,j} - u^n_{i,j} &= \frac{\tau}{2}\left[\delta_x^2\tilde{u}_{i,j} + \delta_y^2 u^n_{i,j}\right]\\ u^{n+1}_{i,j} - \tilde{u}_{i,j} &= \frac{\tau}{2}\left[\delta_x^2\tilde{u}_{i,j} + \delta_y^2 u^{n+1}_{i,j}\right]\end{aligned} \tag{5.71}$$

or

$$\begin{aligned}\left[1 - \tfrac{1}{2}\tau\delta_x^2\right]\tilde{\mathbf{u}} &= \left[1 + \tfrac{1}{2}\tau\delta_y^2\right]\mathbf{u}^n\\ \left[1 - \tfrac{1}{2}\tau\delta_y^2\right]\mathbf{u}^{n+1} &= \left[1 + \tfrac{1}{2}\tau\delta_x^2\right]\tilde{\mathbf{u}}\end{aligned} \tag{5.72}$$

where

$$\mathbf{u}^n = u_{i,j}^n, \qquad \text{for all } i \text{ and } j$$

These formulas were first introduced by Peaceman and Rachford [13], and produce an approximate solution which has an associated error of $0(\Delta t^2 + h^2)$. A higher-accuracy split formula is due to Fairweather and Mitchell [14] and is

$$[1 - \tfrac{1}{2}(\tau - \tfrac{1}{6})\,\delta_x^2]\tilde{\mathbf{u}} = [1 + \tfrac{1}{2}(\tau + \tfrac{1}{6})\,\delta_y^2]\mathbf{u}^n$$

$$[1 - \tfrac{1}{2}(\tau - \tfrac{1}{6})\,\delta_y^2]\mathbf{u}^{n+1} = [1 + \tfrac{1}{2}(\tau + \tfrac{1}{6})\,\delta_x^2]\tilde{\mathbf{u}} \tag{5.73}$$

with an error of $0(\Delta t^2 + h^4)$. Both of these methods are unconditionally stable. A general discussion of ADI methods is given by Douglas and Gunn [15].

The intermediate value $\tilde{\mathbf{u}}$ introduced in each ADI method is not necessarily an approximation to the solution at any time level. As a result, the boundary values at the intermediate level must be chosen with care. If

$$w(x, y, t) = g(x, y, t) \tag{5.74}$$

when (x, y, t) is on the bounadry of the region for which (5.64) is specified, then for (5.72)

$$\tilde{u}_{i,j} = \tfrac{1}{2}(1 - \tfrac{1}{2}\,\tau\delta_y^2)g_{i,j}^{n+1} + \tfrac{1}{2}(1 + \tfrac{1}{2}\,\tau\delta_y^2)\,g_{i,j}^n \tag{5.75}$$

and for (5.73)

$$\tilde{u}_{i,j} = \frac{\tau - \frac{1}{6}}{2\tau}[1 - \tfrac{1}{2}(\tau - \tfrac{1}{6})\,\delta_y^2]g_{i,j}^{n+1} + \frac{\tau + \frac{1}{6}}{2\tau}[1 + \tfrac{1}{2}(\tau + \tfrac{1}{6})\,\delta_y^2]g_{i,j}^n \tag{5.76}$$

If g is not dependent on time, then

$$\tilde{u}_{i,j} = g_{i,j} \qquad \text{(for 5.72)} \tag{5.77}$$

$$\tilde{u}_{i,j} = (1 + \tfrac{1}{6}\,\delta_y^2)g_{i,j} \qquad \text{(for 5.73)} \tag{5.78}$$

A more detailed investigation of intermediate boundary values in ADI methods is given in Fairweather and Mitchell [16].

ADI methods have also been developed for finite element methods. Douglas and Dupont [17] formulated ADI methods for parabolic problems using Galerkin methods, as did Dendy and Fairweather [18]. The discussion of these methods is beyond the scope of this text, and the interested reader is referred to Chapter 6 of [11].

MATHEMATICAL SOFTWARE

As with software for the solution of parabolic PDEs in one space variable and time, the software for solving multidimensional parabolic PDEs uses the method of lines. Thus a computer algorithm for multidimensional parabolic PDEs based upon the MOL must include a spatial discretization routine and a time integrator. The principal obstacle in the development of multidimensional PDE software is the solution of large, sparse matrices. This same problem exists for the development of elliptic PDE software.

Parabolics

The method of lines is used exclusively in these codes. Table 5.1 lists the parabolic PDE software and outlines the type of spatial discretization and time integration for each code. None of the major libraries—NAG, Harwell, and IMSL—contain multidimensional parabolic PDE software, although 2DEPEP is an IMSL product distributed separately from their main library. As with one-dimensional PDE software, the overwhelming choice of the time integrator for multidimensional parabolic PDE software is the Gear algorithm. Next, we illustrate the use of two codes.

Consider the problem of Newtonian fluid flow in a rectangular duct. Initially, the fluid is at rest, and at time equal to zero, a pressure gradient is imposed upon the fluid that causes it to flow. The momentum balance, assuming a constant density and viscosity, is

$$\rho \frac{\partial V}{\partial t} = \frac{P_0 - P_L}{L} + \mu \left[\frac{\partial^2 V}{\partial x^2} + \frac{\partial^2 V}{\partial y^2} \right] \tag{5.79}$$

TABLE 5.1 Parabolic PDE Codes

Code	Spatial Discretization	Time Integrator	Spatial Dimension	Region	Reference
DSS/2	Finite difference	Options including Runge-Kutta and GEARB [24]	2 or 3	Rectangular	[19]
PDETWO	Finite difference	GEARB [24]	2	Rectangular	[20]
FORSIM VI	Finite difference	Options including Runge-Kutta and GEAR [25]	2 or 3	Rectangular	[21]
DISPL	Finite element; Galerkin with tensor products of B-splines for the basis function	Modified version of GEAR [25]	2	Rectangular	[22]
2DEPEP	Finite element; Galerkin with quadratic basis functions on triangular elements; curved boundaries incorporated by isoparametric method	Crank-Nicolson or an implicit method	2	Irregular	[23]

where

$$\rho = \text{fluid density}$$

$$\frac{P_0 - P_L}{L} = \text{pressure gradient}$$

$$\mu = \text{fluid viscosity}$$

$$V = \text{axial fluid velocity}$$

The situation is pictured in Figurc 5.14. Let

$$X = \frac{x}{B}$$

$$Y = \frac{y}{W}$$

$$\eta = \frac{V}{(P_0 - P_L)B^2/(2\mu L)}$$

$$\tau = \frac{\mu t}{\rho B^2} \tag{5.80}$$

Substitution of (5.80) into (5.79) gives

$$\frac{\partial \eta}{\partial \tau} = 2 + \frac{\partial^2 \eta}{\partial^2 X} + \left(\frac{B}{W}\right)^2 \frac{\partial^2 \eta}{\partial^2 Y} \tag{5.81}$$

The subsidiary conditions for (5.81) are

$$\eta = 0 \quad \text{at} \quad \tau = 0 \qquad \text{(fluid initially at rest)}$$

$$\eta = 0 \quad \text{at} \quad Y = 0 \qquad \text{(no slip at the wall)}$$

$$\eta = 0 \quad \text{at} \quad X = 1 \qquad \text{(no slip at the wall)}$$

$$\frac{\partial \eta}{\partial X} = 0 \quad \text{at} \quad X = 0 \qquad \text{(symmetry)}$$

$$\frac{\partial \eta}{\partial Y} = 0 \quad \text{at} \quad Y = 1 \qquad \text{(symmetry)}$$

Equation (5.81) was solved using DISPL (finite element discretization) and PDETWO (finite difference discretization). First let us discuss the numerical results form these codes. Table 5.2 shows the affect of the mesh spacing ($\Delta Y = \Delta X = h$) when solving (5.81) with PDETWO. Since the spatial discretization is accomplished using finite differences, the error associated with this

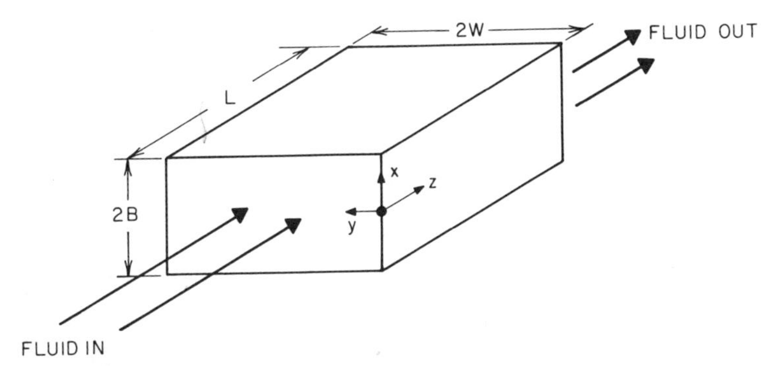

FIGURE 5.14 Flow in a rectangular duct.

discretization is $0(h^2)$. As h is decreased, the values of η shown in Table 5.2 increase slightly. For mesh spacings less than 0.05, the same results were obtained as those shown for $h = 0.05$. Notice that the tolerance on the time integration is 10^{-7}, so the error is dominated by the spatial discretization. When solving (5.81) with DISPL (cubic basis functions), a mesh spacing of $h = 0.25$ produced the same solution as that shown in Table 5.2 ($h = 0.05$). This is an expected result since the finite element discretization is $0(h^4)$.

Figure 5.15 shows the results of (5.81) for various X, Y, and τ. In Figure 5.15*a* the affect at the Y-position upon the velocity profile in the X-direction is illustrated. Since $Y = 0$ is a wall where no slip occurs, the magnitude of the velocity at a given X-position will increase as one moves away from the wall. Figure 5.15*b* shows the transient behavior of the velocity profile at $Y = 1.0$. As one would expect, the velocity increases for $0 \leq X < 1$ as τ increases. This trend would continue until steady state is reached. An interesting question can now be asked. That is, how large must the magnitude of W be in comparison to the magnitude of B to consider the duct as two infinite parallel plates. If the duct in Figure 5.14 represents two infinite parallel plates at $X = \pm 1$, then the

TABLE 5.2 Results of (5.81) Using PDETWO: $\tau = 0.5, \dfrac{B}{W} = 1, Y = 1, \text{TOL} = 10^{-7}$

	η		
X	$h = 0.2$	$h = 0.1$	$h = 0.05$
0.0	0.5284	0.5323	0.5333
0.2	0.5112	0.5149	0.5159
0.4	0.4575	0.4608	0.4617
0.6	0.3614	0.3640	0.3646
0.8	0.2132	0.2146	0.2150
1.0	0	0	0

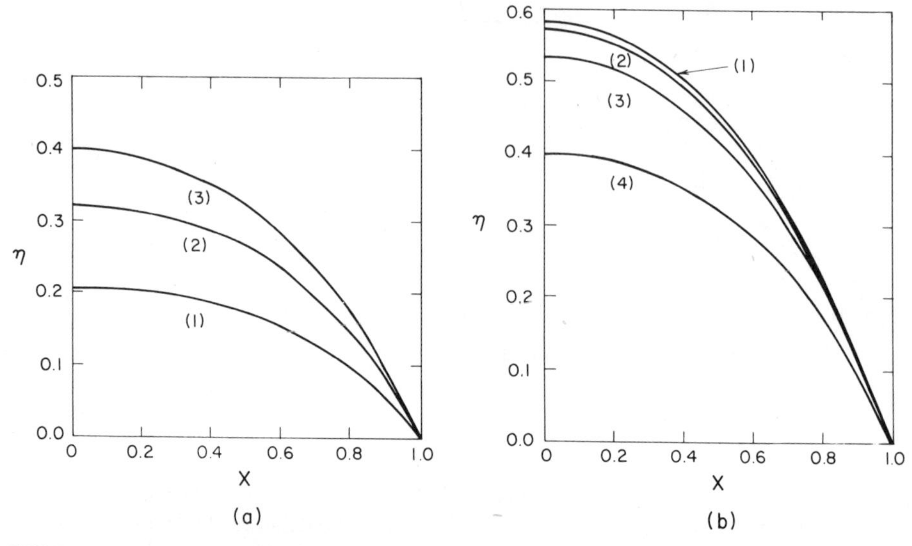

FIGURE 5.15 Results of (5.81).
(*a*) $\tau = 0.25$, $B/W = 1$. (*b*) $Y = 1.0$, $B/W = 1$.

Y
(1) 0.25
(2) 0.5
(3) 1.0

τ
(1) 1.00
(2) 0.75
(3) 0.50
(4) 0.25

momentum balance becomes

$$\frac{\partial \eta}{\partial \tau} = 2 + \frac{\partial^2 \eta}{\partial X^2} \tag{5.82}$$

with

$$\eta = 0 \quad \text{at} \quad \tau = 0$$

$$\eta = 0 \quad \text{at} \quad X = 1$$

$$\frac{\partial \eta}{\partial X} = 0 \quad \text{at} \quad X = 0$$

Equation (5.82) possesses an analytic solution that can be used in answering the posed question. Figure 5.16 shows the affect of the ratio B/W on the velocity profile at various τ. Notice that at low τ, a B/W ratio of $\frac{1}{2}$ approximates the analytical solution of (5.82). At larger τ this behavior is not observed. To match the analytical solution (five significant figures) at all τ, it was found that the value of B/W must be $\frac{1}{4}$ or less.

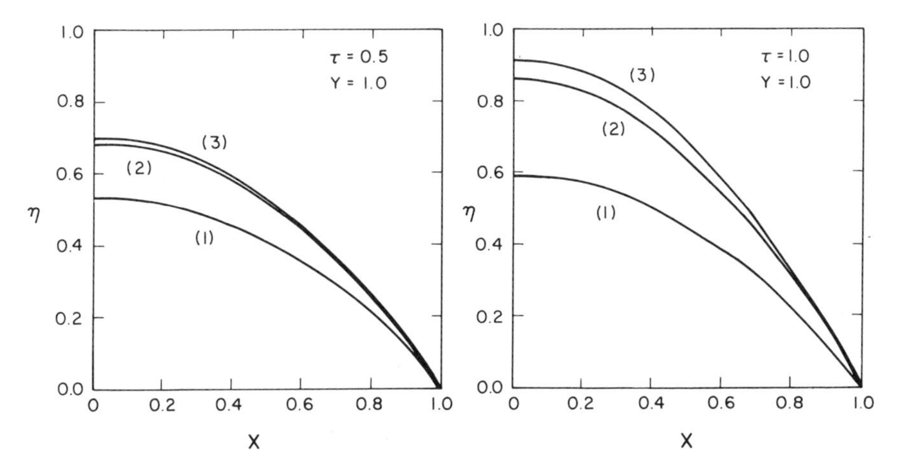

FIGURE 5.16 Further results of (5.81).

B/W

(1) 1
(2) ½
(3) ¼ and analytical solution of (5.82)

Elliptics

Table 5.3 lists the elliptic PDE software and outlines several features of each code. Notice that the NAG library does contain elliptic PDE software, but this routine is not very robust. Besides the software shown in Table 5.3, DISPL and 2DEPEP contain options to solve elliptic PDEs. Next we consider a practical problem involving elliptic PDEs and illustrate the solution and physical implications through the use of DISPL.

The most common geometry of catalyst pellets is the finite cylinder with length to diameter, L/D, ratios from about 0.5 to 4, since they are produced by either pelleting or by extrusion. The governing transport equations for a finite cylindrical catalyst pellet in which a first-order chemical reaction is occurring are [34]:

$$\text{(Mass)} \qquad \frac{\partial^2 f}{\partial r^2} + \frac{1}{r}\frac{\partial f}{\partial r} + \left(\frac{D}{L}\right)^2 \frac{\partial^2 f}{\partial z^2} = \phi^2 f \exp\left[\frac{\gamma}{t}(t-1)\right]$$

$$\text{(Energy)} \qquad \frac{\partial^2 t}{\partial r^2} + \frac{1}{r}\frac{\partial t}{\partial r}\left(\frac{D}{L}\right)^2 \frac{\partial^2 t}{\partial z^2} = -\beta\phi^2 f \exp\left[\frac{\gamma}{t}(t-1)\right] \qquad (5.83)$$

where

r = dimensionless radial coordinate, $0 \leqslant r \leqslant 1$
z = dimensionless axial coordinate, $0 \leqslant z \leqslant 1$

f = dimensionless concentration
t = dimensionless temperature
γ = Arrhenius number (dimensionless)
ϕ = Thiele modulus (dimensionless)
β = Prater number (dimensionless)

with the boundary conditions

$$\frac{\partial f}{\partial r} = \frac{\partial t}{\partial r} = 0 \quad \text{at} \quad r = 0 \qquad \text{(symmetry)}$$

$$\frac{\partial f}{\partial z} = \frac{\partial t}{\partial z} = 0 \quad \text{at} \quad z = 0 \qquad \text{(symmetry)}$$

$$f = t = 1 \quad \text{at} \quad z = 1 \text{ and } r = 1 \qquad \text{(concentration and temperature specified at the surface of the pellet)}$$

Using the Prater relationship [35], which is

$$t = 1 + (1 - f)\beta$$

TABLE 5.3 Elliptic PDE Codes

Code	Discretization	Region	Nonlinear Equations	Reference
NAG (D03 chapter)	Finite difference (Laplace's equation in two dimensions)	Rectangular	No	—
FISPACK	Finite difference	Rectangular	No	[26]
EPDE1	Finite difference	Irregular	No	[27]
ITPACK/REGION	Finite difference	Irregular	No	[28]
FFT9	Finite difference	Irregular	No	[29]
HLMHLZ/HEL-MIT/HELSIX/HELSYM	Finite difference	Irregular	No	[30]
PLTMG	Finite element; Galerkin with linear basis functions on triangular elements	Irregular	No	[31]
ELIPTI	ADI with finite differences; integrate to steady state	Irregular	Yes	[32]
ELLPACK	Finite difference; finite element (collocation and Galerkin)	Rectangular	Yes	[33]

TABLE 5.4 Results of (5.84) Using DISPL

$z = 0.25, \beta = 0.1, \gamma = 30, \frac{D}{L} = 1$

	$\phi = 1$		$\phi = 2$		
r	$h = 0.5$	$h = 0.25$	$h = 0.5$	$h = 0.25$	$h = 0.125$
0	0.728	0.728	0.724(−3)	0.240(−1)	0.227(−1)
0.25	0.745	0.745	0.384(−1)	0.377(−1)	0.365(−1)
0.50	0.797	0.797	0.109	0.115	0.115
0.75	0.882	0.882	0.414	0.404	0.404
1.0	1.000	1.000	1.000	1.000	1.000

reduces the system (5.83) to the single elliptic PDE:

$$\frac{\partial^2 f}{\partial r^2} + \frac{1}{r}\frac{\partial f}{\partial r} + \left(\frac{D}{L}\right)^2 \frac{\partial^2 f}{\partial z^2} = \phi^2 f \exp\left[\frac{\gamma\beta(1-f)}{1+\beta(1-f)}\right] \tag{5.84}$$

$$\frac{\partial f}{\partial r} = 0 \quad \text{at} \quad r = 0$$

$$\frac{\partial f}{\partial z} = 0 \quad \text{at} \quad z = 0$$

$$f = 1 \quad \text{at} \quad r = 1 \text{ and } z = 1$$

DISPL (using cubic basis functions) produced the results given in Tables 5.4 and 5.5 and Figure 5.17. In Table 5.4 the affect of the mesh spacing ($h = \Delta r = \Delta z$) is shown. With $\phi = 1$ a coarse mesh spacing ($h = 0.5$) is sufficient to give three-significant-figure accuracy. At larger values of ϕ a finer mesh is required for a similar accuracy. As ϕ increases, the gradient in f becomes larger, especially near the surface of the pellet. This behavior is shown in Figure 5.17. Because of this gradient, a finer mesh is required to obtain an accurate solution over the entire region. Alternatively, one could refine the mesh in the region of the steep gradient. Finally, in Table 5.5 the isothermal results ($\beta = 0$) are compared with those published elsewhere [34]. As shown, DISPL produced accurate results with $h = 0.25$.

TABLE 5.5 Further Results of (5.84) Using DISPL

$\beta = 0.0, \gamma = 30, \phi = 3, \frac{L}{D} = 1$

(r, z)	DISPL, $h = 0.25$	From Reference [34]
(0.394, 0.285)	0.337	0.337
(0.394, 0.765)	0.585	0.585
(0.803, 0.285)	0.648	0.648
(0.803, 0.765)	0.756	0.759

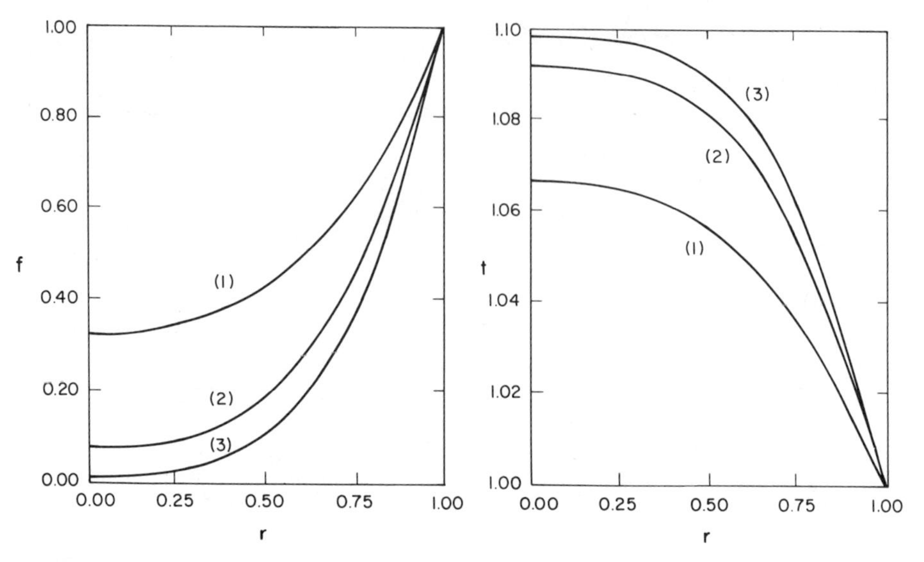

FIGURE 5.17 Results of (5.84): $\beta = 0.1$, $\gamma = 30$, $\phi = 2$, $D/L = 1$.

z
(1) 0.75
(2) 0.50
(3) 0.00

PROBLEMS

1. Show that the finite difference discretization of

$$(x + 1)\frac{\partial^2 w}{\partial x^2} + (y^2 + 1)\frac{\partial^2 w}{\partial y^2} - w = 1$$

$$0 \leq x \leq 1, \qquad 0 \leq y \leq 1, \qquad \Delta x = \Delta y = \tfrac{1}{3}$$

with

$$w(0, y) = y$$
$$w(1, y) = y^2$$
$$w(x, 0) = 0$$
$$w(x, 1) = 1$$

is given by [36]:

$$\begin{bmatrix} 5 & -\frac{10}{9} & -\frac{4}{3} & 0 \\ -\frac{13}{9} & \frac{17}{3} & 0 & -\frac{4}{3} \\ -\frac{5}{3} & 0 & \frac{17}{3} & -\frac{10}{9} \\ 0 & -\frac{5}{3} & -\frac{13}{9} & \frac{19}{3} \end{bmatrix} \begin{bmatrix} u_{1,1} \\ u_{1,2} \\ u_{2,1} \\ u_{2,2} \end{bmatrix} = \begin{bmatrix} \frac{1}{3} \\ \frac{20}{9} \\ \frac{2}{27} \\ \frac{56}{27} \end{bmatrix}$$

2.* Consider a rectangular plate with an initial temperature distribution of

$$w(x, y, 0) = T - T_0 = 0, \qquad 0 \leq x \leq 2, \quad 0 \leq y \leq 1$$

If the edges $x = 2$, $y = 0$, and $y = 1$ are held at $T = T_0$ and on the edge $x = 0$ we impose the following temperature distribution:

$$w(0, y, t) = T - T_0 = \begin{cases} 2ty, & \text{for } 0 \leq y \leq \frac{1}{2} \\ 2t(1 - y), & \text{for } \frac{1}{2} \leq y \leq 1 \end{cases}$$

solve the heat conduction equation

$$\frac{\partial w}{\partial t} = \frac{\partial^2 w}{\partial x^2} + \frac{\partial^2 w}{\partial y^2}$$

for the temperature distribution in the plate. The analytical solution to this problem is [22]:

$$w = \frac{4}{\pi} \sum_{m=1}^{\infty} \sum_{n=1}^{\infty} \frac{m}{n^2} \frac{1}{\sigma^2} (e^{-\sigma t} + \sigma t - 1) \sin\left(\frac{n\pi}{2}\right) \sin\left(\frac{m\pi x}{2}\right) \sin(n\pi y)$$

where

$$\sigma = \pi^2 \left(\frac{m^2}{4} + n^2\right)$$

Calculate the error in the numerical solution at the mesh points.

3.* An axially dispersed isothermal chemical reactor can be described by the following material balance equation:

$$\frac{\partial f}{\partial z} = \frac{1}{\text{Pe}_r}\left[\frac{\partial^2 f}{\partial r^2} + \frac{1}{r}\frac{\partial f}{\partial r}\right] + \frac{1}{\text{Pe}_a}\frac{\partial^2 f}{\partial z^2} + D_a f, \qquad 0 \leq r \leq 1, \quad 0 \leq z \leq 1$$

with

$$1 - f = \frac{1}{\text{Pe}_a}\frac{\partial f}{\partial z} \quad \text{at} \quad z = 0, \qquad \frac{\partial f}{\partial r} = 0 \quad \text{at} \quad r = 0 \text{ and } r = 1$$

$$\frac{\partial f}{\partial z} = 0 \quad \text{at} \quad z = 1$$

where

f = dimensionless concentration

r = dimensionless radial coordinate

z = dimensionless axial coordinate

Pe_r = radial Peclet number

Pe_a = axial Peclet number

D_a = Damkohler number (first-order reaction rate)

The boundary conditions in the axial direction arise from continuity of flux as discussed in Chapter 1 of [34]. Let $D_a = 0.5$ and $Pe_r = 10$. Solve the material balance equation using various values of Pe_a. Compare your results to plug flow ($Pe_a \to \infty$) and discuss the effects of axial dispersion.

4.* Solve Eq. (5.84) with $D/L = 1$, $\phi = 1$, $\gamma = 30$, and let $-0.2 \leq \beta \leq 0.2$. Comment on the affect of varying β [$\beta < 0$ (endothermic), $\beta > 0$ (exothermic)].

5.* Consider transient flow in a rectangular duct, which can be described by:

$$\frac{\partial \eta}{\partial \tau} = \overline{\alpha} + \frac{\partial^2 \eta}{\partial X^2} + \left(\frac{B}{W}\right)^2 \frac{\partial^2 \eta}{\partial Y^2}$$

using the same notation as with Eq. (5.81) where $\overline{\alpha}$ is a constant. Solve the above equation with

	$\overline{\alpha}$	Comment
(a)	2	Eq. (5.81)
(b)	4	Twice the pressure gradient as Eq. (5.81)
(c)	1	Half the pressure gradient as Eq. (5.81)

How does the pressure gradient affect the time required to reach steady state?

REFERENCES

1. Bank, R. E., and D. J. Rose, "Parameter Selection for Newton-Like Methods Applicable to Nonlinear Partial Differential Equations: SIAM J. Numer. Anal., *17,* 806 (1980).
2. Denison, K. S., C. E. Hamrin, and J. C. Diaz, "The Use of Preconditioned Conjugate Gradient and Newton-Like Methods for a Two-Dimensional, Nonlinear, Steady-State, Diffusion, Reaction Problem," Comput. Chem. Eng., *6,* 189 (1982).
3. Denison, K. S., private communication (1982).
4. Varga, R. S., *Matrix Iterative Analysis,* Prentice-Hall, Englewood Cliffs, N.J. (1962).
5. Birkhoff, G., M. H. Schultz, and R. S. Varga, "Piecewise Hermite Interpolation in One and Two Variables with Applications to Partial Differential Equations," Numer. Math., *11,* 232 (1968).
6. Bramble, J. H., and S. R. Hilbert, "Bounds for a Class of Linear Functionals with Applications to Hermite Interpolation," Numer. Math., *16,* 362 (1971).

7. Hilbert, S. R., "A Mollifier Useful for Approximations in Sobolev Spaces and Some Applications to Approximating Solutions of Differential Equations," Math. Comput., *27,* 81 (1973).
8. Prenter, P. M., and R. D. Russell, "Orthogonal Collocation for Elliptic Partial Differential Equations," SIAM J. Numer. Anal., *13,* 923 (1976).
9. Courant, R., "Variational Methods for the Solution of Problems of Equilibrium and Vibrations," Bull. Am. Math. Soc., *49,* 1 (1943).
10. Mitchell, A. R., and R. Wait, *The Finite Element Method in Partial Differential Equations,* Wiley, London (1977).
11. Fairweather, G., *Finite Element Galerkin Methods for Differential Equations,* Marcel Dekker, New York (1978).
12. Strang, G., and G. J. Fix, *An Analysis of the Finite Element Method,* Prentice-Hall, Englewood Cliffs, N.J. (1973).
13. Peaceman, D. W., and H. H. Rachford, "The Numerical Solution of Parabolic and Elliptic Differential Equations," J. Soc. Ind. Appl. Math., *3,* 28 (1955).
14. Mitchell, A. R., and G. Fairweather, "Improved Forms of the Alternating Direction Methods of Douglas, Peaceman and Rachford for Solving Parabolic and Elliptic Equations," Numer. Math. *6,* 285 (1964).
15. Douglas, J., and J. E. Gunn, "A General Formulation of Alternating Direction Methods. Part I. Parabolic and Hyperbolic Problems," Numer. Math., *6,* 428 (1964).
16. Fairweather, G., and A. R. Mitchell, "A New Computational Procedure for A.D.I. Methods," SIAM J. Numer. Anal., 163 (1967).
17. Douglas, J. Jr., and T. Dupont, "Alternating Direction Galerkin Methods on Rectangles," in *Numerical Solution of Partial Differential Equations II,* B. Hubbard (ed.), Academic, New York (1971).
18. Dendy, J. E. Jr., and G. Fairweather, "Alternating-Direction Galerkin Methods for Parabolic and Hyperbolic Problems on Rectangular Polygons," SIAM J. Numer. Anal., *12,* 144 (1975).
19. Schiesser, W., "DSS/2—An Introduction to the Numerical Methods of Lines Integration of Partial Differential Equations," Lehigh Univ., Bethlehem, Pa. (1976).
20. Melgaard, D., and R. Sincovec, "General Software for Two Dimensional Nonlinear Partial Differential Equations," ACM TOMS, *7,* 106 (1981).
21. Carver, M., et al., "The FORSIM VI Simulation Package for the Automated Solution of Arbitrarily Defined Partial Differential and/or Ordinary Differential Equation Systems, Rep. AECL-5821, Chalk River Nuclear Lab., Ontario, Canada (1978).
22. Leaf, G. K., M. Minkoff, G. D. Byrne, D. Sorensen, T. Bleakney, and J. Saltzman, "DISPL: A Software Package for One and Two Spatially

Dimensional Kinetics-Diffusion Problems," Rep. ANL-77-12, Argonne National Lab., Argonne, Ill. (1977).

23. Sewell, G., "A Finite Element Program with Automatic User-Controlled Mesh Grading," in *Advances in Computer Methods for Partial Differential Equations III,* R. Vishnevetsky and R. S. Stepleman (eds.), IMACS (AICA), Rutgers Univ., New Brunswick, N.J. (1979).

24. Hindmarsh, A. C., "GEARB: Solution of Ordinary Differential Equations Having Banded Jacobians," Lawrence Livermore Laboratory, Report UCID-30059 (1975).

25. Hindmarsh, A. C., "GEAR: Ordinary Differential Equation System Solver," Lawrence Livermore Laboratory Report UCID-30001 (1974).

26. Adams, J., P. Swarztrauber, and N. Sweet, "FISHPAK: Efficient FORTRAN Subprograms for the Solution of Separable Elliptic Partial Differential Equations: Ver. 3, Nat. Center Atmospheric Res., Boulder, Colo. (1978).

27. Hornsby, J., "EPDE1—A Computer Programme for Elliptic Partial Differential Equations (Potential Problems)," Computer Center Program Library Long Write-Up D-300, CERN, Geneva (1977).

28. Kincaid, D., and R. Grimes, "Numerical Studies of Several Adaptice Iterative Algorithms," Report 126, Center for Numerical Analysis, Univ. Texas, Austin (1977).

29. Houstics, E. N., and T. S. Papatheodorou, "Algorithm 543. FFT9, Fast Solution of Helmholtz-Type Partial Differential Equations," ACM TOMS, *5,* 490 (1979).

30. Proskurowski, W., "Four FORTRAN Programs for Numerically Solving Helmholtz's Equation in an Arbitrary Bounded Planar Region," Lawrence Berkeley Laboratory Report 7516 (1978).

31. Bank, R. E., and A. H. Sherman, "PLTMG Users' Guide," Report CNA 152, Center for Numerical Analysis, Univ. Texas, Austin (1979).

32. Taylor, J. C., and J. V. Taylor, "ELIPTI-TORMAC: A Code for the Solution of General Nonlinear Elliptic Problems over 2-D Regions of Arbitrary Shape," in *Advances in Computer Methods for Partial Differential Equations II,* R. Vichnevetsky (ed.), IMACS (AICA), Rutgers Univ., New Brunswick, N.J. (1977).

33. Rice, J., "ELLPACK: A Research Tool for Elliptic Partial Differential Equation Software," in *Mathematical Software III,* J. Rice (ed.), Academic, New York (1977).

34. Villadsen, J., and M. L. Michelsen, *Solution of Differential Equation Models by Polynomial Approximation,* Prentice-Hall, Englewood Cliffs, N.J. (1978).

35. Prater, C. D., "The Temperature Produced by Heat of Reaction in the Interior of Porous Particles," Chem. Eng. Sci., *8,* 284 (1958).

36. Ames, W. F., *Numerical Methods for Partial Differential Equations,* 2nd ed., Academic, New York (1977).

37. Dixon, A. G., "Solution of Packed-Bed Heat-Exchanger Models by Orthogonal Collocation Using Piecewise Cubic Hermite Functions," MCR Tech. Summary Report #2116, University of Wisconsin-Madison (1980).

BIBLIOGRAPHY

An overview of finite difference and finite element methods for partial differential equations in several space variables has been given in this chapter. For additional or more detailed information, see the following texts:

Finite Difference

Ames, W. F., *Nonlinear Partial Differential Equations in Engineering,* Academic, New York (1965).

Ames, W. F. (ed.), *Nonlinear Partial Differential Equations,* Academic, New York (1967).

Ames, W. F., *Numerical Methods for Partial Differential Equations,* 2nd ed., Academic, New York (1977).

Finlayson, B. A., *Nonlinear Analysis in Chemical Engineering,* McGraw-Hill, New York (1980).

Mitchell, A. R., and D. F. Griffiths, *The Finite Difference Method in Partial Differential Equations,* Wiley, Chichester (1980).

Finite Element

Becker, E. B., G. F. Carey, and J. T. Oden, *Finite Elements: An Introduction,* Prentice-Hall, Englewood Cliffs, N.J. (1981).

Fairweather, G., *Finite Element Galerkin Methods for Differential Equations,* Marcel Dekker, New York (1978).

Huebner, K. H., *The Finite Element Method for Engineers,* Wiley, New York (1975).

Mitchell, A. R., and D. F. Griffiths, *The Finite Difference Method in Partial Differential Equations,* Wiley, Chichester (1980). Chapter 5 discusses the Galerkin method.

Mitchell, A. R., and R. Wait, *The Finite Element Method in Partial Differential Equations,* Wiley, New York (1977).

Strang, G., and G. J. Fix, *An Analysis of the Finite Element Method,* Prentice-Hall, Englewood Cliffs, N.J. (1973).

Zienkiewicz, O. C., *The Finite Element Method in Engineering Science,* McGraw-Hill, New York (1971).

APPENDIX

A Computer Arithmetic and Error Control

In mathematical computations on a computer, errors are introduced into the solutions. These errors are brought into a calculation in three ways:

1. Error is present at the outset in the original data—*inherent error*
2. Error results from replacing an infinite process by a finite one—*truncation error,* i.e., representing a function by the first few terms of a Taylor series expansion
3. Error arises as a result of the finite precision of the numbers that can be represented in a computer—*round-off error.*

Each of these errors is unavoidable in a calculation, and hence the problem is not to prevent their occurrence, but rather to control their magnitude. The control of inherent error is not within the scope of this text, and the truncation errors pertaining to specific methods are discussed in the appropriate chapters. This section outlines computer arithmetic and how it influences round-off errors.

COMPUTER NUMBER SYSTEM

The mathematician or engineer, in seeking a solution to a problem, assumes that all calculations will be performed within the system of real numbers, $\mathscr{R}$. In $\mathscr{R}$, the interval between any two real numbers contains infinitely many real numbers. $\mathscr{R}$ does not exist in a computer because there are only a finite amount of real numbers within a computer's number system. This is a source of round-off error. In computer memory, each number is stored in a location that consists of a sign (±) plus a fixed number of digits. A discussion of how these digits represent numbers is presented next.

NORMALIZED FLOATING-POINT NUMBER SYSTEM

A floating-point number system is characterized by four parameters:

$$\beta = \text{number base}$$
$$\text{t} = \text{precision}$$
$$L, U = \text{exponent range.}$$

One can denote such a system by

$$F(\beta, t, L, U)$$

Each floating-point number, $x \neq 0$, in F is represented in the following way:

$$x = \pm \left(\frac{d_1}{\beta} + \frac{d_2}{\beta^2} + \ldots + \frac{d_t}{\beta^t} \right) \times \beta^e \qquad \textbf{(A.1)}$$

where

$$1 \leq d_1 < \beta$$
$$0 \leq d_s < \beta, \qquad 2 \leq s \leq t,$$
$$L \leq e \leq U$$

The fact that $d_1 \neq 0$ means that the floating-point number system is normalized.

ROUND-OFF ERRORS

Next, consider the differences between computations in F versus $\mathcal{R}$, i.e., round-off errors. The source of the differences lies in the fact that F is not closed under the arithmetic operations of addition and multiplication (likewise, subtraction and division); the sum or the product of two numbers in F may not necessarily be an element of F. Hence, to stay in F, the computer replaces the "true" result of an operation by an element of F, and this process produces some error. Several cases can occur [A.4]:

1. The exponent e of the result can lie outside the range $L \leq e \leq U$,
(a) If $e > U$, *overflow;* for example, in $F(2, 3, -1, 2)$ **(A.2)**

$$(0.100 \times 2^2) \times (0.110 \times 2^2) = 0.110 \times 2^{\underline{3}}$$
$$2 \qquad \times \qquad 3 \qquad = \qquad 6$$

(b) If $e < L$, *underflow;* for example, in $F(2, 3, -1, 2)$ **(A.3)**

$$(0.100 \times 2^0) \times (0.110 \times 2^{-1}) = 0.110 \times 2^{\underline{2}}$$
$$\frac{1}{2} \quad \times \quad \frac{3}{8} \quad = \quad \frac{3}{16}$$

2. The fractional part has more than t digits; for example, consider $F(2, 3, -1, 2)$

$$(0.110 \times 2^0) + (0.111 \times 2^0) = 0.1101 \times 2^1$$
$$\frac{3}{4} \quad + \quad \frac{7}{8} \quad = \quad \frac{13}{8}$$

(notice that four digits are required to represent the fractional part). Similarly,

$$(0.111 \times 2^0) \times (0.110 \times 2^0) = 0.10101 \times 2^0$$
$$\frac{7}{8} \quad \times \quad \frac{3}{4} \quad = \quad \frac{21}{32}$$

(while this situation does not arise frequently in addition, it almost invariably does with multiplication).

To define a result that can be represented in the machine, the computer selects a nearby element of F. This can be done in two ways: rounding and chopping. Suppose the "true" result of an operation is

$$\left(\frac{d_1}{\beta} + \frac{d_2}{\beta^2} + \ldots + \frac{d_t}{\beta^t} + \frac{d_{t+1}}{\beta^{t+1}} + \ldots \frac{d_w}{\beta^w}\right) \times \beta^e \qquad \textbf{(A.4)}$$

then,

1. Chopping: digits beyond $(d_t)/(\beta^t)$ are dropped.
2. Rounding:
$$\left(\frac{d_1}{\beta} + \frac{d_2}{\beta^2} + \ldots + \frac{d_{t+1} + \frac{1}{2}\beta}{\beta^{t+1}}\right) \times \beta^e$$
then chop.

For example, if one considers $F(2, 3, -1, 2)$, the number

$$0.1101 \times 2^1 \rightarrow 0.110 \times 2^1\text{: chopping}$$
$$0.1101 \times 2^1 \rightarrow 0.111 \times 2^1\text{: rounding,}$$

while for

$$0.10101 \times 2^0 \rightarrow 0.101 \times 2^0\text{: chopping}$$
$$0.10101 \times 2^0 \rightarrow 0.101 \times 2^0\text{: rounding.}$$

Both methods are commonly used on present-day computers. No matter the method, there is some round-off error introduced by the process. If $f(x)$ represents the machine representation of x, then

$$\sigma(x) = \text{relative round-off error} = \left|\frac{x - f(x)}{x}\right|, \qquad x \neq 0$$

It can be shown that [A.1]

$$\sigma(x) \leq \text{EPS} = \begin{cases} \beta^{1-t}: & \text{chopping} \\ \frac{1}{2}\beta^{2-t}: & \text{rounding} \end{cases} \tag{A.5}$$

As an example, suppose $x = 12.467$ with $F(10, 4, -50, 50)$ and chopping, then $f(x) = 0.1246 \times 10^2$ and

$$\sigma(x) = \left|\frac{12.467 - 0.1246 \times 10^2}{12.467}\right|$$

or

$$\sigma(x) = 0.00056 < \text{EPS} = 10^{-3}$$

For the same system with rounding, $f(x) = 0.1247 \times 10^{-2}$ and

$$\sigma(x) = 0.00024 < \text{EPS} = \tfrac{1}{2} \times 10^{-3}$$

One can see that the parameter EPS plays an important role in computation with a floating-point number system. EPS is the machine epsilon and is defined to be the smallest positive machine number such that

$$f(1 + \text{EPS}) > 1,$$

For example, for $F(10, 4, -50, 50)$ with chopping

$$\text{EPS} = 10^{-3} \quad \text{since} \quad f(1 + 0.001) = 0.1001 \times 10^1 > 1$$

and for rounding

$$\text{EPS} = 0.0005 \quad \text{since} \quad f(1 + 0.0005) = 0.1001 \times 10^1 > 1$$

The machine epsilon is an indicator of the attainable accuracy in a floating-point number system and can be used to determine the maximum achievable accuracy of computation.

Take, as a specific example, an IBM 3032 and find EPS. Considering only floating-point number systems, the IBM 3032 uses either of two base 16 systems:

1. $F_s(16, 6, -64, 63)$: single precision
2. $F_D(16, 14, -64, 63)$: extended precision

For chopping ($\text{EPS} = \beta^{1-t}$):

$$\text{EPS (single)} = 9.54 \times 10^{-7}$$

$$\text{EPS (extended)} = 2.22 \times 10^{-16}$$

If one executes the following algorithm (from Forsythe, Malcolm, and Moler [A.1]):

```
      DOUBLE PRECISION EPS, EPS1
      EPS = 1.D0
   10 EPS = 0.5D0*EPS
      EPS1 = EPS + 1.D0
      IF (EPS1. GT. 1.D0) GO TO 10
      WRITE (6,20) EPS
   20 FORMAT (5X, 'THE MACHINE EPSILON = ', D17.10)
      STOP
      END
```

the result is:

```
THE MACHINE EPSILON = 0.1110223025 D-15
```

This method of finding EPS can differ from the "true" EPS by at most a fraction of 2 (EPS is continually halved in statement number 10). Notice that the calculated value of EPS is half of the value predicted by EPS $= \beta^{1-t}$, as one would expect. In the course of carrying out a computer calculation of practical magnitude, a very large number of arithmetic operations are performed, and the errors can propagate. It is, therefore, wise to use the number system with the greatest precision.

Another computational problem involving the inability of the computer to represent numbers of $\mathscr{R}$ in F is shown below. Take for instance the number 0.1, which is used frequently in the partition of intervals, and consider whether ten steps of length 0.1 are the same as one step of length 1.0. If one executes the following algorithm on an IBM 3032:

```
      DOUBLE PRECISION X
      X = 0.D0
      N = 0
      DO 10 I = 1,10
      X = X + 0.1D0
   10 CONTINUE
      IF (X.EQ.1.D0) N = 1
      WRITE(6,20) N,X
   20 FORMAT (10X,' THE VALUE OF N = ',I1,/,10X,
     ★' THE VALUE OF X = ',D17.10)
      STOP
      END
```

the result is:

```
THE VALUE OF N = 0
THE VALUE OF X = 0.1000000000D 01.
```

Since the printed value of x is exactly 1.0, then why is the value of N still equal to zero? The answer to this question is as follows. The IBM computer operates with β being a power of 2, and because of this, the number 0.1 cannot be exactly represented in F (0.1 does not have a terminating expansion of $\frac{1}{2}$). In fact,

$$\frac{1}{10} = \frac{0}{2} + \frac{0}{2^2} + \frac{0}{2^3} + \frac{1}{2^4} + \frac{1}{2^5} + \frac{0}{2^6} + \frac{0}{2^7} + \ldots$$

or

$$(0.1)_{10} = (0.000110011001100\ldots)_2 = (0.19999\ldots)_{16}$$

The base 2 or base 16 representations are terminated after t digits since the IBM chops when performing computations, and when ten of these representations of 0.1 are added together, the result is not exactly 1.0. This is why N was not set equal to 1 in the above algorithm. Why then is the printed value of x equal to 1.0? The IBM machine chops when performing computations, but then rounds on output. Therefore, it is the rounding procedure on output that sets x exactly equal to 1.0.

The programmer must be aware of the subtleties discussed in this appendix and many others, which are described in Chapter 2 of [A.1], for effective implementation of computational algorithms.

APPENDIX

B
Newton's Method

Systems of nonlinear algebraic equations arise in the discretization of differential equations. In this appendix, we illustrate a technique for solving systems of nonlinear algebraic equations. More detailed discussions of this topic can be found in [A.1–A.4].

Consider the set of nonlinear algebraic equations

$$
\begin{aligned}
f_1(y_1, y_2, \ldots, y_n) &= 0 \\
f_2(y_1, y_2, \ldots, y_n) &= 0 \\
&\vdots \\
f_n(y_1, y_2, \ldots, y_n) &= 0
\end{aligned}
\qquad \textbf{(B.1)}
$$

which can be written as

$$f_i(\mathbf{y}) = 0, \qquad i = 1, 2, \ldots, n,$$

or

$$\mathbf{f}(\mathbf{y}) = \mathbf{0}$$

We wish to find that set $\{y_i | i = 1, \ldots, n\}$ that satisfies (B.1).

Although there are many ways to solve Eq. (B.1), the most common method of practical use is Newton's method (or variants of it). In the case of a single equation, the Newton method consists in linearizing the given equation $f(y) = 0$ by approximating $f(y)$ by

$$f(y^0) + f'(y^0)(y - y^0) \qquad \textbf{(B.2)}$$

where y^0 is believed to be close to the actual solution, and solving the linearized equation

$$f(y^0) + f'(y^0)\,\Delta y = 0 \tag{B.3}$$

The value $y^1 = y^0 + \Delta y$ is then accepted as a better approximation, and the process is continued if necessary.

Now consider the system (B.1). If the ith equation is linearized, then

$$f_i(y_1^k, y_2^k, \ldots, y_n^k) + \sum_{j=1}^{n} \left[\left.\frac{\partial f_i}{\partial y_j}\right|_k (y_j^{k+1} - y_j^k) \right] = 0 \tag{B.4}$$

where $k \geqslant 0$. The Jacobian is defined as

$$J_{ij}^k = \left.\frac{\partial f_i}{\partial y_j}\right|_k \tag{B.5}$$

and (B.4) can be written in matrix form as

$$J^k\,\Delta\mathbf{y} = -\mathbf{f}(\mathbf{y}^k) \tag{B.6}$$

where

$$\Delta y_j = y_j^{k+1} - y_j^k$$

$$\Delta\mathbf{y} = (\Delta y_1, \Delta y_2, \ldots, \Delta y_n)^T$$

The procedure is

1. Choose y^0
2. Calculate $\Delta\mathbf{y}$ from (B.6)
3. Set $\mathbf{y}^{k+1} = \mathbf{y}^k + \Delta\mathbf{y}$
4. Iterate on (2) and (3) until

   $$\|\Delta\mathbf{y}\|_\infty < \text{TOL}$$

   where

   $$\|\mathbf{x}\|_\infty = \max x_i$$

   $$\text{TOL} = \text{arbitrary}$$

The convergence of the Newton method is proven in [A.2] under certain conditions, and it is shown that the method converges quadratically, i.e.,

$$\|\mathbf{g}^* - \mathbf{y}^{k+1}\|_\infty < m\,\|\mathbf{g}^* - \mathbf{y}^k\|_\infty^2 \tag{B.7}$$

where

$$\mathbf{f}(\mathbf{g}^*) = \mathbf{0}$$

and

$$m = \text{a constant}$$

APPENDIX C

Gaussian Elimination

From the main body of this text one can see that all the methods for solving differential equations can yield large sets of equations that can be formulated into a matrix problem. Normally, these equations give rise to a matrix having a special property in that a great many of its elements are zero. Such matrices are called sparse. Typically, there is a pattern of zero and nonzero elements, and special matrix methods have been developed to take these patterns into consideration. In this appendix we begin by discussing a method for solving a general linear system to equations and then proceed by outlining a method for the tridiagonal matrix.

DENSE MATRIX

The standard method of solving a linear system of algebraic equations is to do a lower-upper (LU) decomposition on the matrix, or Gaussian elimination. Consider a dense (all elements are nonzero), nonsingular (all rows or columns are independent) $n \times n$ matrix $\boldsymbol{A}$ such that

$$\boldsymbol{A}\mathbf{x} = \mathbf{r}$$

where

$$\mathbf{x} = [x_1, x_2, \ldots, x_n]^T$$

$$\mathbf{r} = [r_1, r_2, \ldots, r_n]^T$$

$$A = \begin{bmatrix} a_{11} & a_{12} & \cdots & a_{1n} \\ a_{21} & & & \cdot \\ \cdot & & & \cdot \\ \cdot & & & \cdot \\ \cdot & \cdot & \cdot & a_{nn} \end{bmatrix}$$

The 21 element can be made zero by multiplying the first row by $-a_{21}/a_{11}$ and adding it to the second row. By multiplying the first row by $-a_{31}/a_{11}$ and adding to the third row, the 31 element becomes zero. Likewise,

$$A^{[1]}\mathbf{x} = \begin{bmatrix} a_{11} & a_{12} & a_{13} & \cdots \\ 0 & a_{22} - \dfrac{a_{21}}{a_{11}} a_{12} & a_{23} - \dfrac{a_{21}}{a_{11}} a_{13} & \cdots \\ 0 & a_{32} - \dfrac{a_{31}}{a_{11}} a_{12} & a_{33} - \dfrac{a_{31}}{a_{11}} a_{13} & \cdots \\ \cdot & \cdot & \cdot & \cdot \\ \cdot & \cdot & \cdot & \cdot \\ \cdot & \cdot & \cdot & \cdot \end{bmatrix} \begin{bmatrix} x_1 \\ x_2 \\ \cdot \\ \cdot \\ \cdot \\ x_n \end{bmatrix} = \begin{bmatrix} r_1 \\ r_2 - \dfrac{a_{21}}{a_{11}} r_1 \\ r_3 - \dfrac{a_{31}}{a_{11}} r_1 \\ \cdot \\ \cdot \\ \cdot \end{bmatrix} = \mathbf{r}^{[1]} \tag{C.2}$$

In sequel this is

$$a_{i,j}^{[k]} = a_{i,j}^{[k-1]} - \frac{a_{i,k-1}^{[k-1]}}{a_{k-1,k-1}^{[k-1]}} a_{k-1,j}^{[k-1]}, \tag{C.3}$$

$$r_i^{[k]} = r_i^{[k-1]} - \frac{a_{i,k-1}^{[k-1]}}{a_{k-1,k-1}^{[k-1]}} r_{k-1} \tag{C.4}$$

Now make a column of zeros below the diagonal in the second column by doing the same process as before

$$A^{[2]}\mathbf{x} = \begin{bmatrix} a_{11} & a_{12} & a_{13} & \cdots \\ 0 & a_{22}^{[2]} & a_{23}^{[2]} & \cdots \\ 0 & 0 & a_{33}^{[2]} & \cdots \\ \cdot & \cdot & a_{43}^{[2]} & \cdots \\ \cdot & \cdot & \cdot & \cdots \\ \cdot & \cdot & \cdot & \cdots \\ \cdot & \cdot & \cdot & \cdots \end{bmatrix} \begin{bmatrix} x_1 \\ x_2 \\ x_3 \\ \cdot \\ \cdot \\ \cdot \\ \cdot \end{bmatrix} = \begin{bmatrix} r_1 \\ r_2^{[2]} \\ r_3^{[2]} \\ \cdot \\ \cdot \\ \cdot \\ \cdot \end{bmatrix} \tag{C.5}$$

Continue the procedure until the lower triangle is filled with zeros and set $\boldsymbol{A}^{[n]} = \boldsymbol{U}$.

$$\boldsymbol{U}\mathbf{x} = \boldsymbol{A}^{[n]}\mathbf{x} = \begin{bmatrix} a_{11} & a_{12} & a_{13} & a_{14} & \cdots \\ & a_{22}^{[2]} & a_{23}^{[2]} & a_{24}^{[2]} & \cdots \\ & & a_{33}^{[3]} & a_{34}^{[3]} & \cdots \\ & & & a_{44}^{[4]} & \cdots \\ & 0 & & & \cdots \\ & & & & \cdots \\ & & & & \cdots \end{bmatrix} \begin{bmatrix} x_1 \\ x_2 \\ \cdot \\ \cdot \\ \cdot \\ \cdot \\ \cdot \end{bmatrix} = \begin{bmatrix} r_1 \\ r_2^{[2]} \\ r_3^{[3]} \\ \cdot \\ \cdot \\ \cdot \\ \cdot \end{bmatrix} \tag{C.6}$$

Define $\boldsymbol{L}$ as the matrix with zeros in the upper triangle, ones on the diagonal, and the scalar multiples used in the lower triangle to create $\boldsymbol{U}$,

$$\boldsymbol{L} = \begin{bmatrix} 1 & & & & & \\ -\dfrac{a_{21}}{a_{11}} & 1 & & 0 & & \\ -\dfrac{a_{31}}{a_{11}} & -\dfrac{a_{32}^{[2]}}{a_{22}^{[2]}} & 1 & \cdot & & \\ \cdot & \cdot & & & \cdot & \\ \cdot & \cdot & & & & \cdot \\ \cdot & \cdot & & -\dfrac{a_{n,n-1}^{[n-1]}}{a_{n-1,n-1}^{[n-1]}} & & 1 \end{bmatrix} \tag{C.7}$$

If the unit diagonal is understood, then $\boldsymbol{L}$ and $\boldsymbol{U}$ can be stored in the same space as $\boldsymbol{A}$. The solution is now obtained by

$$\begin{aligned} x_n &= \frac{r_n^{[n]}}{a_{nn}^{[n]}} \\ x_{n-1} &= \frac{r_{n-1}^{[n-1]} - a_{n-1,n}^{[n-1]}\, x_n}{a_{n-1,n-1}^{[n-1]}} \\ x_i &= \frac{r_i^{[i]} - \sum_{j=i+1}^{n} a_{i,j}^{[i]}\, x_j}{a_{i,i}^{[i]}} \end{aligned} \tag{C.8}$$

It is possible to show that $\boldsymbol{A} = \boldsymbol{LU}$ [A.5]. Thus (C.1) can be represented as

$$\boldsymbol{A}\mathbf{x} = \boldsymbol{LU}\mathbf{x} = \mathbf{r} \tag{C.9}$$

Notice that the sequence

$$Ly = \mathbf{r} \tag{C.10}$$

$$U\mathbf{x} = \mathbf{y} \tag{C.11}$$

gives (C.9) by multiplying (C.11) from the left by L to give

$$LU\mathbf{x} = L\mathbf{y} \tag{C.12}$$

or

$$A\mathbf{x} = L\,U\,\mathbf{x} = \mathbf{r}$$

One can think of Eq. (C.3) as the LU decomposition of A, Eq. (C.4) as the forward substitution or the solution of (C.10), and Eq. (C.8) as the backward elimination or the solution of (C.11). If a certain problem has a constant matrix A but different right-hand sides $\mathbf{r}$, then the matrix need only be decomposed once and the iteration would only involve the forward and backward sweeps.

The number of multiplications and divisions needed to do one LU decomposition and m-forward and backward sweeps for a dense matrix is

$$\text{OP}_{\text{G.E.}} = \tfrac{1}{3}n^3 - \tfrac{1}{3}n + mn^2 \tag{C.13}$$

This is fewer operations than it takes to calculate an inverse, so the decomposition is more efficient. Notice that the decomposition is proportional to n^3, whereas the forward and backward sweeps are proportional to n^2. For large n the decomposition is a significant cost.

The only way that Gaussian elimination can become unstable and the process break down when A is nonsingular is if $a_{ii}^{[i-1]} = 0$ before performing step i of the decomposition. Since the procedure is being performed on a computer, round-off errors can cause $a_{ii}^{[i-1]}$ to be "close" to zero, likewise, causing instabilities. Often this round-off error problem can be avoided by pivoting; that is, find row s such that $\max\limits_{i \leq j \leq n} |a_{ji}^{[i-1]}| = a_{si}^{[i-1]}$ and switch row s and row i before performing the ith step. To avoid pivoting, we must impose that matrix A be diagonally dominant:

$$|a_{ii}| \geq \sum_{\substack{j=1 \\ j \neq i}}^{n} |a_{ij}|, \qquad i = 1, \ldots, n, \tag{C.14}$$

where the strict inequality must hold for at least one row. Condition (C.14) insures that $a_{ii}^{[i-1]}$ will not be "close" to zero, and therefore the Gaussian elimination procedure is stable and does not require pivoting.

TRIDIAGONAL MATRIX

The LU decomposition of a tridiagonal matrix is performed by Gaussian elimination. A tridiagonal matrix can be written as

$$\begin{bmatrix} b_1 & c_1 & & & & & \\ a_2 & b_2 & c_2 & & & & \\ & a_3 & b_3 & c_3 & & & \\ & & \cdot & \cdot & \cdot & & \\ & & & \cdot & \cdot & \cdot & \\ & & & \cdot & \cdot & \cdot & \\ & & & a_{n-1} & b_{n-1} & c_{n-1} \\ & & & & a_n & b_n \end{bmatrix} \begin{bmatrix} x_1 \\ x_2 \\ \cdot \\ \\ \cdot \\ \\ \cdot \\ x_n \end{bmatrix} = \begin{bmatrix} r_1 \\ r_2 \\ \cdot \\ \\ \cdot \\ \\ \cdot \\ r_n \end{bmatrix} \quad \text{(C.15)}$$

The Thomas algorithm (Gaussian elimination which takes into account the form of the matrix) is

$$\alpha_1 = \frac{c_1}{b_1} \quad \text{(C.16)}$$

$$\gamma_1 = \frac{r_1}{b_1} \quad \text{(C.17)}$$

$$\alpha_i = \frac{c_i}{b_i - a_i\alpha_{i-1}}, \qquad i = 2, 3, \ldots, n \quad \text{(C.18)}$$

$$\gamma_i = \frac{r_i - a_i\gamma_{i-1}}{b_i - a_i\alpha_{i-1}}, \qquad i = 2, 3, \ldots, n \quad \text{(C.19)}$$

with

$$c_n = 0,$$

and

$$x_n = \gamma_n \quad \text{(C.20)}$$

$$x_i = \gamma_i - \alpha_i x_{i+1}, \qquad i = n - 1, n - 2, \ldots, 1 \quad \text{(C.21)}$$

Equations (C.18) and (C.19) are the LU decomposition and forward substitution, and Eq. (C.21) is the backward elimination. The important point is that there is no fill outside the tridiagonal matrix (structure remains the same). This is an advantage in reducing the work and storage requirements. The operation count to solve m such systems of size n is

$$\mathrm{OP}_{\mathrm{TD}} = 2(n - 1) + m(3n - 2),$$

which is a significant savings over $(\frac{1}{3})n^3$ of (C.13). Since this algorithm is a special form of Gaussian elimination without pivoting, the procedure is stable only when the matrix possesses diagonal dominance.

APPENDIX D
B-Splines

In finite element methods the manner in which the approximate numerical solution of a differential equation is represented affects the entire solution process. Specifically, we would like to choose an approximating space of functions that is easy to work with and is capable of approximating the solution accurately. Such spaces exist, and bases for these spaces can be constructed by using B-splines [A.6]. The authoritative text on this subject is by deBoor [A.6].

Before defining the B-splines, one must first understand the meaning of a divided difference and a truncated power function. The first-order divided difference of a function $g(x)$, $x_i \leq x \leq x_{i+1}$, is

$$g[x_i, x_{i+1}] = \frac{g(x_{i+1}) - g(x_i)}{x_{i+1} - x_i} \tag{D.1}$$

while the higher-order divided difference (dd) formulas are given by recursion formulas: the rth-order dd of $g(x)$ on the points $x_i, x_{i+1}, \ldots, x_{i+r}$ is

$$g[x_i, x_{i+1}, \ldots, x_{i+r}] = \frac{g[x_{i+1}, \ldots, x_{i+r}] - g[x_i, \ldots, x_{i+r-1}]}{x_{i+r} - x_i} \tag{D.2}$$

where

$$g[x_i, \ldots, x_{i+r-1}] = \frac{g[x_{i+1}, \ldots, x_{i+r-1}] - g[x_i, \ldots, x_{i+r-2}]}{x_{i+r-1} - x_i}$$

$$\vdots \qquad\qquad \vdots$$

$$g[x_i, x_{i+1}, x_{i+2}] = \frac{g[x_{i+1}, x_{i+2}] - g[x_i, x_{i+1}]}{x_{i+2} - x_i}$$

Notice that the ith-order dd of a function is equal to its ith derivative at some point times a constant. Next, define a truncated power function of order r (degree $= r - 1$) as:

$$\phi_x^r(t) = (x - t)_+^{r-1} = \begin{cases} (x - t)^{r-1}, & x \geqslant t \\ 0, & x < t \end{cases} \qquad \textbf{(D.3)}$$

This function is illustrated in Figure D.1. The function and all of its derivatives except for the $(r - 1)$st are continuous [$(r - 1)$th derivative is not continuous at $x = t$]. Now, let the sequence $t_j, \ldots, t_{j+r}$ of $r + 1$ points be a nondecreasing sequence and define

$$Z_j^r(x) = \phi_x^r[t_j, \ldots, t_{j+r}] \qquad \textbf{(D.4)}$$

Thus, $Z_j^r(x) = 0$, when $t_j, \ldots, t_{j+r}$ is not in an interval containing x, and when the interval does contain x, $Z_j^r(x)$ is a linear combination of terms $\phi_x^r(t) = (x - t)_+^{r-1}$ evaluated at $t = t_j, \ldots, t_{j+r}$, that is, the linear combination that results from the dd's.

The ith B-spline of order k for the sequence $\mathbf{t} = \{t_i | \, i = 1, \ldots, k\}$ (called a knot sequence) is denoted by $B_{i,k,\mathbf{t}}$ and is defined by

$$B_{i,k,\mathbf{t}}(x) = (-1)^k (t_{i+k} - t_i) Z_i^k(x) \qquad \textbf{(D.5)}$$

If k and $\mathbf{t}$ are understood, then one can write $\mathrm{B}_i(x)$ for $B_{i,k,\mathbf{t}}(x)$. The main properties of $B_i(x)$ are:

1. Each $B_i(x) = 0$ when $x < t_i$ or $x > t_{i+k}$ (local support).
2. $\sum_{\text{all } i} B_i(x) = 1$, specifically $\sum_{i=q+1-k}^{s} B_i(x) = 1$ for $t_q \leqslant x \leqslant t_s$.
3. Each $B_i(x)$ satisfied $0 \leqslant B_i(x) \leqslant 1$ for $t_i \leqslant x \leqslant t_{i+k}$ [normalized by the term $(t_{i+k} - t_i)$ in (D.5)] and possesses only one maximum.

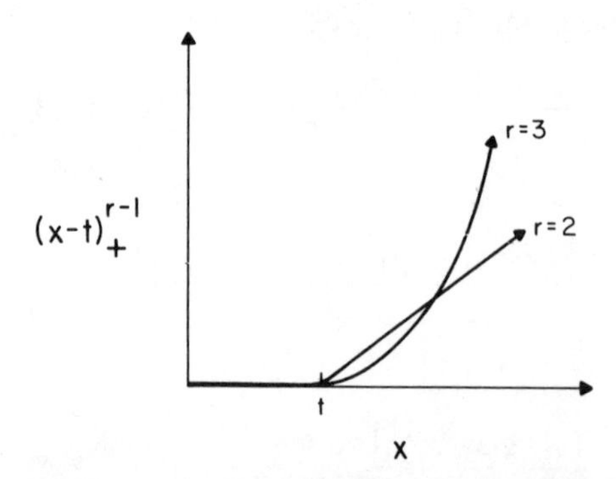

FIGURE D.1 Truncated power function of order *r*.

Consider the approximating space $\mathscr{P}_k^{\boldsymbol{\nu}}(\pi)$ (described in Chapter 3) and whether the B-splines can serve as a basis for this space. Given the partition π,

$$a = x_1 < x_2 < \ldots < x_{\ell+1} = b, \tag{D.6}$$

and the nonnegative integer sequence $\boldsymbol{\nu} = \{\nu_j | j = 2, \ldots, \ell\}$, which denotes the continuity at the breakpoints x_i, then the dimension of $\mathscr{P}_k^{\boldsymbol{\nu}}(\pi)$ is given by

$$N = \dim \mathscr{P}_k^{\boldsymbol{\nu}}(\pi) = \sum_{j=1}^{\ell} (k - \nu_j) \tag{D.7}$$

with $\nu_1 = 0$. If $\mathbf{t} = \{t_i | i = 1, \ldots, N + k\}$ such that

$t_1 \leqslant t_2 \leqslant \ldots \leqslant t_k \leqslant x_1$ (makes the first B-spline one at x_1)

$x_{\ell+1} \leqslant t_{N+1} \leqslant \ldots \leqslant t_{N+k}$ (makes the last B-spline one at $x_{\ell+1}$)

and if for $i = 2, \ldots, \ell$, the number x_i occurs exactly $k - \nu_i$ times in the set $\mathbf{t}$, then the sequence $B_i(x)$, $i = 1, \ldots, N$ is a basis for $\mathscr{P}_k^{\boldsymbol{\nu}}(\pi)$ [A.6]. Therefore a function $f(x)$ can be represented in $\mathscr{P}_k^{\boldsymbol{\nu}}(\pi)$ by

$$f(x) = \sum_{i=1}^{N} \alpha_i B_i(x) \tag{D.9}$$

The B-splines have many computational features that are described in [A.6]. For example, they are easy to evaluate. To evaluate a B-spline at a point x, $t_j \leqslant x \leqslant t_{j+1}$, the following algorithm can be used [A.7] [let $B_{i,k,\mathbf{t}}(x)$ be denoted by $B_{i,k}$ and $Z_i^k(x)$ by $Z_{i,k}$]:

```
      B(i,1) = 1
      DO 20 ℓ = 1, . . ., k - 1
        B(i-ℓ,ℓ+1) = 0
        DO 10 j = 1, . . ., ℓ
          Z(i+j-ℓ,ℓ) = B(i+j-ℓ,ℓ)/(t(i+j) - t(i+1-ℓ))
          B(i+j-ℓ-1,ℓ+1) = B(i+j-ℓ-1,ℓ+1) + (t(i+j) - x)Z(i+j-ℓ,ℓ)
          B(i+j-ℓ,ℓ+1) = (x - t(i+j-ℓ))Z(i+j-ℓ,ℓ)
10      CONTINUE
20    CONTINUE
```

Thus B-splines of lower order are used to evaluate B-splines of higher order. A complete set of algorithms for computing with B-splines is given by deBoor [A.6].

A B-spline $B_{i,k,\mathbf{t}}(x)$ has support over $[t_i, \ldots, t_{i+k}]$. If each point x_i appears only once in $\mathbf{t}$, that is, $\nu_i = k - 1$, then the support is k subintervals, and the B-spline has continuity of the $k - 2$ derivative and all lower-order derivatives.

To decrease the continuity, one must increase the number of times x_i appears in **t** (this also decreases the support). This loss in continuity results from the loss in order of the dd's. To illustrate this point, consider the case of quadratic B-splines ($k = 3$) corresponding to the knots {0, 1, 1, 3, 4, 6, 6, 6} on the partition $x_1 = 1, x_2 = 3, x_3 = 4, x_4 = 6$. For this case notice that ($k = 3$, $\ell = 3$):

$$\nu_2 = 2, \qquad \nu_3 = 2$$

$$N = \dim \mathscr{P}_k^{\nu}(\pi) = \sum_{j=1}^{3} (3 - \nu_j) = 5$$

$$t_1 \leqslant \ldots \leqslant t_k \leqslant x_1 \qquad (0 \leqslant 1 \leqslant 1)$$

and

$$x_{\ell+1} \leqslant t_{N+1} \leqslant \ldots \leqslant t_{N+k} \qquad (6 \leqslant 6 \leqslant 6)$$

Therefore, there are five B-splines to be calculated, each of order 3. Figure D.2 (from [A.6]) illustrates these B-splines. From this figure one can see the normal parabolic spline, $B_3(x)$, which comes from the fact that all the knots are distinct. Also, the loss of continuity in the first derivative is illustrated, for example, $B_2(x)$ at $x = 1$ due to the repetition of x_1.

When using B-splines as the basis functions for finite element methods, one specifies the order k and the knot sequence. From this information, one can calculate the basis if continuity in all derivatives of order lower than the order of the differential equation is assumed.

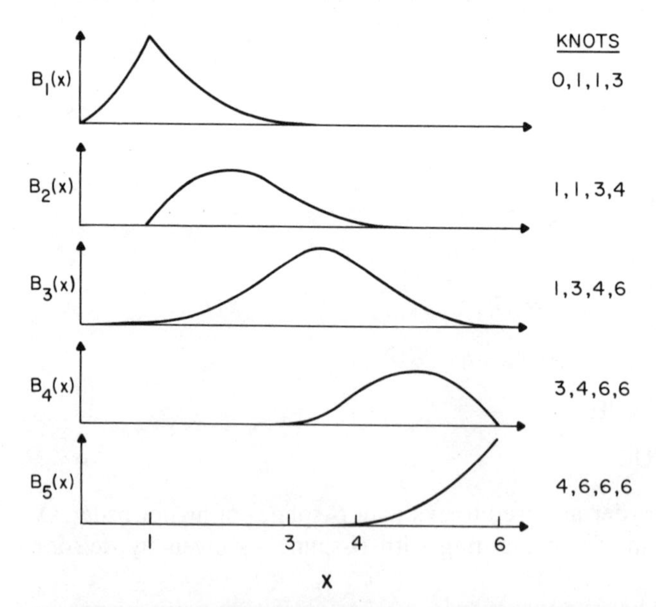

FIGURE D.2 Illustration of B-Splines. Adapted from Carl de Boor, *A Practical Guide to Splines*, copyright © 1978, p. 112. Reprinted by permission of Springer-Verlag, Heidelberg, and the author.

APPENDIX E

Iterative Matrix Methods

Consider the solution of the linear algebraic system

$$\boldsymbol{A}\mathbf{x} = \mathbf{b} \tag{E.1}$$

where $\boldsymbol{A}$ is a given real $N \times N$ matrix, $\mathbf{b}$ is a given N component vector, and $\mathbf{x}$ is the unknown vector to be determined. In this appendix we are concerned with systems in which N is large and $\boldsymbol{A}$ is sparse. Linear systems of this type arise from the numerical solution of partial differential equations.

After discretizing a partial differential equation by either finite difference or finite element techniques, one is faced with the task of solving a large sparse system of linear equations of form (E.1). Two general procedures used to solve systems of this nature are direct and iterative methods. Direct methods are those that, in the absence of round-off errors, yield the exact solution in a finite number of arithmetic operations. An example of a direct method is Gaussian elimination. Iterative methods are those that start with an initial guess for the solution and that, by applying a suitably chosen algorithm, lead to better approximations. In general, iterative methods require less storage and fewer arithmetic operations than direct methods for large sparse systems (for a comparison of direct and iterative methods, see [A.8]).

An iterative method for the solution of linear systems is obtained by splitting the matrix $\boldsymbol{A}$ into two parts, say

$$\boldsymbol{A} = \boldsymbol{S} - \boldsymbol{T} \tag{E.2}$$

To solve (E.1) define a sequence of vectors $\mathbf{x}^{\ell}$ by

$$\boldsymbol{S}\mathbf{x}^{\ell+1} = \boldsymbol{T}\mathbf{x}^{\ell} + \mathbf{b}, \qquad \ell = 0, 1, \ldots \tag{E.3}$$

where $\mathbf{x}^0$ is specified. If the sequence of vectors converges, then the limit will be a solution of (E.1).

There are three common splittings of $\boldsymbol{A}$. The first is known as the point Jacobi method and is

$$\begin{aligned} \boldsymbol{S} &= \boldsymbol{D} \\ \boldsymbol{T} &= \boldsymbol{D} - \boldsymbol{A} \end{aligned} \tag{E.4}$$

where the matrix $\boldsymbol{D}$ is the diagonal matrix whose main diagonal is that of $\boldsymbol{A}$. In component form the point Jacobi method is

$$x_i^{\ell+1} = -\sum_{\substack{j=1 \\ j\neq i}}^{N} \frac{a_{ij}}{a_{ii}} x_j^{\ell} + \frac{b_i}{a_{ii}}, \qquad 1 \leqslant i \leqslant N, \quad \ell \geqslant 0 \tag{E.5}$$

An obvious necessary condition for (E.5) to work is $a_{ii} \neq 0$. If $\boldsymbol{A}$ is diagonally dominant, then the point Jacobi method converges [A.3]. Examination of (E.5) shows that one must save all the components of $\mathbf{x}^{\ell}$ while computing $\mathbf{x}^{\ell+1}$. The Gauss-Seidel method does not possess this storage requirement. The matrix splitting equations for the Gauss-Seidel method are

$$\begin{aligned} \boldsymbol{S} &= \boldsymbol{D} + \boldsymbol{L} \\ \boldsymbol{T} &= -\boldsymbol{U} \end{aligned} \tag{E.6}$$

where $\boldsymbol{D}$ is as before, and $\boldsymbol{U}$ and $\boldsymbol{L}$ are strictly upper and lower triangular matrices, respectively. In component form this method is

$$x_i^{\ell+1} = -\sum_{j=1}^{i-1} \frac{a_{ij}}{a_{ii}} x_j^{\ell+1} - \sum_{j=i+1}^{N} \frac{a_{ij}}{a_{ii}} x_j^{\ell} + \frac{b_i}{a_{ii}}, \qquad 1 \leqslant i \leqslant N, \quad \ell \geqslant 0. \tag{E.7}$$

As with the point Jacobi method, $\boldsymbol{A}$ must be diagonally dominant for the Gauss-Seidel method to converge [A.3]. Also, in most practical problems the Gauss-Seidel method converges faster than the point Jacobi method. The third common method is closely related to the Gauss-Seidel method. Let the vector $\hat{\mathbf{x}}^{\ell+1}$ be defined by

$$\hat{x}_i^{\ell+1} = -\sum_{j=1}^{i-1} \frac{a_{ij}}{a_{ii}} x_j^{\ell+1} - \sum_{j=i+1}^{N} \frac{a_{ij}}{a_{ii}} x_j^{\ell} + \frac{b_i}{a_{ii}}, \qquad 1 \leqslant i \leqslant N, \quad \ell \geqslant 0 \tag{E.8}$$

from which $\mathbf{x}^{\ell+1}$ is obtained as

$$x_i^{\ell+1} = x_i^{\ell} + \omega(\hat{x}_i^{\ell+1} - x_i^{\ell})$$

or

$$x_i^{\ell+1} = (1 - \omega)x_i^{\ell} + \omega\hat{x}_i^{\ell+1} \tag{E.9}$$

The constant ω, $1 \leqslant \omega \leqslant 2$, is called the relaxation parameter, and is chosen to accelerate the convergence. Equations (E.8) and (E.9) can be combined to give

$$x_i^{\ell+1} = x_i^{\ell} + \omega\left\{-\sum_{j=1}^{i-1} \frac{a_{ij}}{a_{ii}} x_j^{\ell+1} - \sum_{j=i+1}^{N} \frac{a_{ij}}{a_{ii}} x_j^{\ell} + \frac{b_i}{a_{ii}} - x_i^{\ell}\right\}$$

$$1 \leqslant i \leqslant N, \quad \ell \geqslant 0 \tag{E.10}$$

Notice that if $\omega = 1$, the method is the Gauss-Seidel method. Equation (E.10) can be written in the split matrix notation as

$$\begin{aligned} \boldsymbol{S} &= \frac{1}{\omega}[\boldsymbol{D} + \omega \boldsymbol{L}] \\ \boldsymbol{T} &= \frac{1}{\omega}[(1 - \omega)\boldsymbol{D} - \omega \boldsymbol{U}] \end{aligned} \tag{E.11}$$

where $\boldsymbol{D}$, $\boldsymbol{L}$, and $\boldsymbol{U}$ are as previously defined. This method is called successive over relaxation (SOR). In the practical use of SOR, finding the optimal ω is of major importance. Adaptive procedures have been developed for the automatic determination of ω as the iterative procedure is being carried out (see, for example [A.9]).

Computer packages are available for the solution of large, sparse linear systems of algebraic equations. One package, ITPACK [A.10], contains research-oriented programs that implement iterative methods.

APPENDIX REFERENCES

A.1. Forsythe, G. E., M. A. Malcolm, and C. B. Moler, *Computer Methods for Mathematical Computations,* Prentice-Hall, Englewood Cliffs, N.J. (1977).

A.2. Keller, H. B., *Numerical Solution of Two Point Boundary Value Problems,* SIAM, Philadelphia (1976).

A.3. Finlayson, B. A., *Nonlinear Analysis in Chemical Engineering,* McGraw-Hill, New York (1980).

A.4. Johnston, R. L., *Numerical Methods—A Software Approach,* Wiley, New York (1982).

A.5. Forsythe, G., and G. B. Moler, *Computer Solution of Linear Algebraic Systems,* Prentice-Hall, Englewood Cliffs, N.J. (1967).

A.6. deBoor, C., *Practical Guide to Splines,* Springer-Verlag, New York (1978).

A.7. Russell, R. D., *Numerical Solution of Boundary Value Problems,* Lecture Notes, Universidad Central de Venezuela, Publication 79-06, Caracas (1979).

A.8. Eisenstat, S., A. George, R. Grimes, D. Kincaid, and A. Sherman, "Some Comparisons of Software Packages for Large Sparse Linear Systems," in *Advances in Computer Methods for Partial Differential Equations III,* R. Vichneveshy and R. S. Stepleman (eds.), IMACS (AICA), Rutgers University, New Brunswick, N.J. (1979).

A.9. Kincaid, D. R., "On Complex Second-Degree Iterative Methods," SIAM J. Numer. Anal., *2*, 211 (1974).

A.10. Grimes, R. G., D. R. Kincaid, W. I. MacGregor, and D. M. Young, "ITPACK Report: Adaptive Iterative Algorithms Using Symmetric Sparse Storage," Report No. CNA-139, Center for Numerical Analysis, Univ. of Texas, Austin, Tex. (1978).

Author Index

Subject Index